THE FOURTEENERS
Colorado's Great Mountains

THE FOURTEENERS
Colorado's Great Mountains

Perry Eberhart and Philip Schmuck

SAGE BOOKS

THE SWALLOW PRESS INC.
CHICAGO

Copyright © 1970 by Perry Eberhart and Philip Schmuck

All Rights Reserved
Printed in the United States of America

Published by
The Swallow Press Incorporated
1139 South Wabash Avenue
Chicago, Illinois 60605

LIBRARY OF CONGRESS CATALOG CARD NUMBER 72-75740

Map on facing page from Guide to the Colorado Mountains by Robert Ormes; used by permission of the Colorado Mountain Club.

Colorado Mountain Ranges

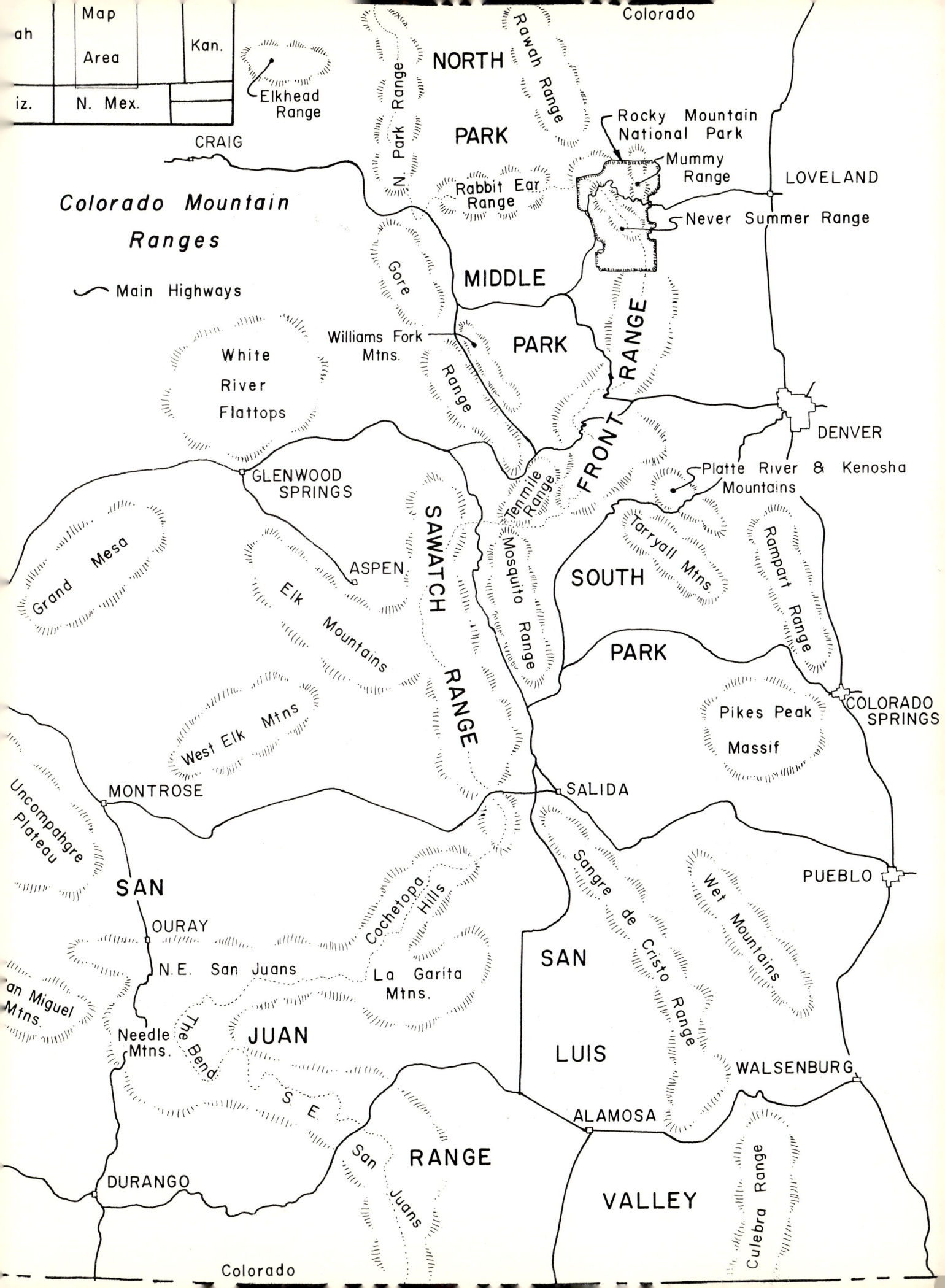

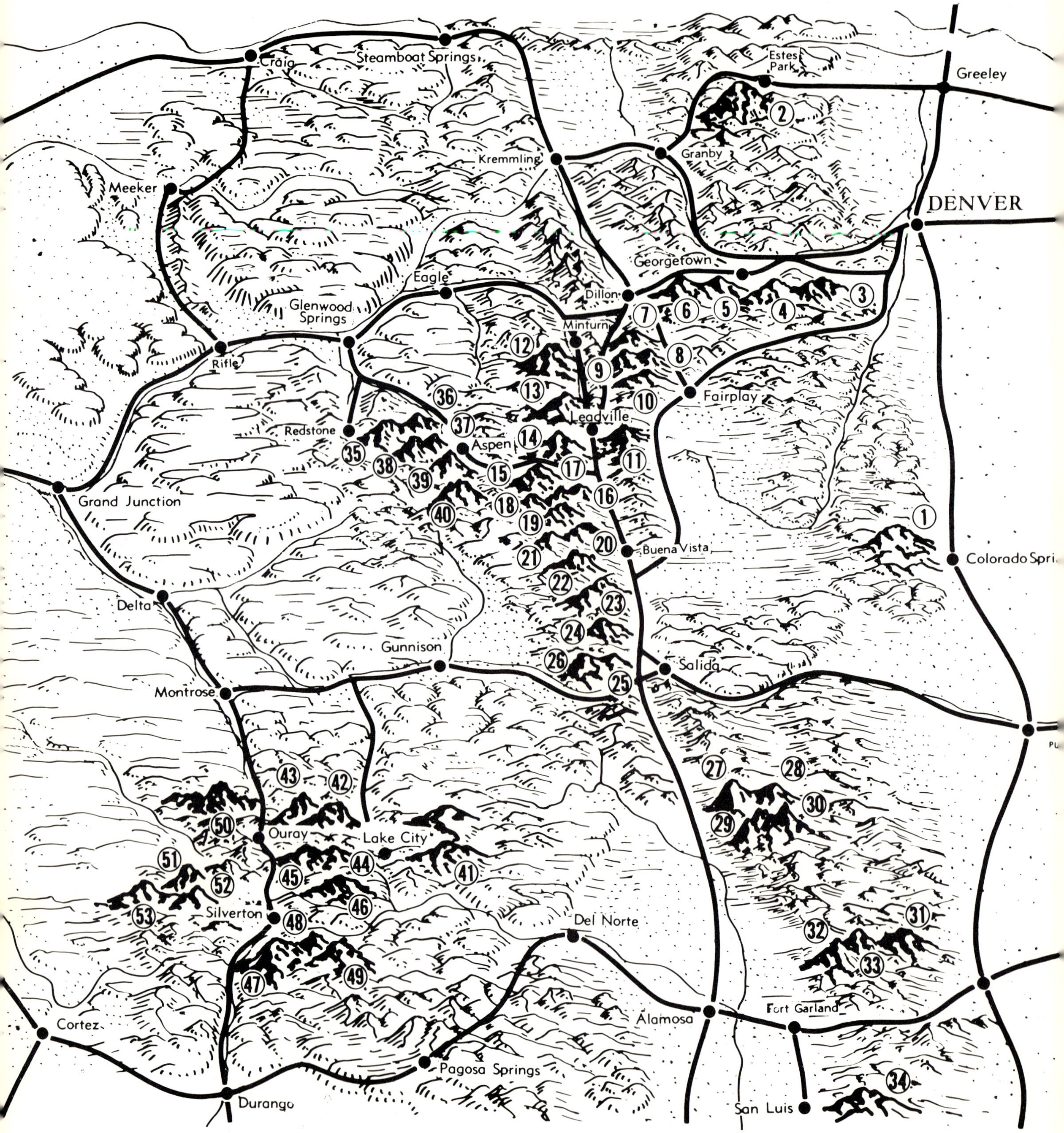

Craig
Steamboat Springs
Estes Park
Greeley
Granby
Kremmling
Meeker
DENVER
Georgetown
Eagle
Dillon
Glenwood Springs
Minturn
Rifle
Fairplay
Leadville
Redstone
Aspen
Grand Junction
Buena Vista
Colorado Spri
Delta
Gunnison
Salida
Montrose
Ouray
Lake City
Silverton
Del Norte
Alamosa
Fort Garland
Cortez
Pagosa Springs
Durango
San Luis
1
2
3
4
5
6
7
8
9
10
11
12
13
14
15
16
17
18
19
20
21
22
23
24
25
26
27
28
29
30
31
32
33
34
35
36
37
38
39
40
41
42
43
44
45
46
47
48
49
50
51
52
53

CONTENTS

		page
	By Way of Introduction	1
1	Pikes Peak	5
2	Longs Peak	8
3	Mount Evans	12
4	Mount Bierstadt	14
5	Grays Peak	16
6	Torreys Peak	18
7	Quandary Peak	20
8	Mount Lincoln	22
9	Mount Democrat	24
10	Mount Bross	28
11	Mount Sherman	30
12	Mount of the Holy Cross	33
13	Mount Massive	36
14	Mount Elbert	38
15	La Plata Peak	41
16	Mount Oxford	42
17	Mount Belford	44
18	Huron Peak	46
19	Missouri Mountain	50
20	Mount Harvard	52
21	Mount Columbia	54
22	Mount Yale	56
23	Mount Princeton	58
24	Mount Antero	60
25	Mount Shavano	62
26	Tabeguache Mountain	64
27	Kit Carson Peak	69

		page
28	Humboldt Peak	70
29	Crestone Peak	73
30	Crestone Needle	76
31	Mount Lindsey	78
32	Little Bear Peak	80
33	Blanca Peak	84
34	Culebra Peak	86
35	Snowmass Peak	89
36	Capitol Peak	90
37	North Maroon Peak	94
38	Maroon Peak or South Maroon Peak	94
39	Pyramid Peak	96
40	Castle Peak	99
41	San Luis Peak	100
42	Uncompahgre Peak	103
43	Wetterhorn Peak	105
44	Sunshine Peak	106
45	Redcloud Peak	108
46	Handies Peak	110
47	Mount Eolus	112
48	Sunlight Peak	114
49	Windom Peak	116
50	Mount Sneffels	120
51	Wilson Peak	122
52	Mount Wilson	124
53	El Diente Peak	126
	Welcome to the Club	128
	Honorable Mention	128

ALPHABETICAL LISTING OF THE FOURTEENERS

	page		page
Antero	60	Longs	8
Belford	44	Maroon	94
Bierstadt	14	Massive	36
Blanca	84	Missouri	50
Bross	28	North Maroon	94
Capitol	90	Oxford	42
Castle	99	Pikes	5
Columbia	54	Princeton	58
Crestone (Needle)	76	Pyramid	96
Crestone (Peak)	73	Quandary	20
Culebra	86	Redcloud	108
Democrat	24	San Luis	100
Elbert	38	Shavano	62
El Diente	126	Sherman	30
Eolus	112	Sneffels	120
Evans	12	Snowmass	89
Grays	16	South Maroon	94
Handies	110	Sunlight	114
Harvard	52	Sunshine	106
Holy Cross	33	Tabeguache	64
Humboldt	70	Torreys	18
Huron	46	Uncompahgre	103
Kit Carson	69	Wetterhorn	105
La Plata	41	Wilson (Mount)	124
Lincoln	22	Wilson (Peak)	122
Lindsey	78	Windom	116
Little Bear	80	Yale	56

BY WAY OF INTRODUCTION

"Because it is there." The beckoning challenge of the mountain works for the non-mountain climber as well. The ever-changing faces of nature's magnificent giants are there for the looking, the marveling at, the wondering about, the worshipping beneath, as well as the climbing. No man can help being awed by them, humbled by them, struck by their beauty.

This book is dedicated to all, the mountain climber, the would-be poet, the casual traveler. Perhaps, in some small way, the book will bring you into closer communion with Colorado's mountain miracles through a better understanding and deeper appreciation of them.

A "Fourteener" is a mountain rising 14,000 feet or higher above sea level. Of some 80 Fourteeners in the U.S. (excluding Alaska), 53 are in Colorado. The number has varied over the years as new measurements were made, as new measurement methods were used, and as new peaks were located or separated from other nearby peaks and given an identity of their own. In 1923 there were 46 Colorado mountains recognized as Fourteeners; 48 in 1925, 50 in 1929, and 52 from the mid-30's to the mid-50's. From then until the mid-60's there were deemed to be 54. Now the number is down to 53.

This book is about all 53 of Colorado's towering personalities. To better understand them, let us remind ourselves of a few basic facts about mountains in general.

Mountains are defined as a land form group composed of rugged areas with crests that are in general more than 2,000 feet above the adjacent lowlands. They are commonly distinguished from other major relief features by their predominance of slope and their overall massiveness. In terms of local relief, mountains are low when they are less than 3,000 feet and high when above 3,000 feet.

Colorado has, by far, more high mountains than any other state (excluding Alaska): there are more than one thousand over 10,000 feet. The altitude of the land forms is generally from one to two miles above sea level; thus the actual height of the highest mountains, from their base, varies from about 4,000 feet to 7,000 feet.

The Rocky Mountains are composed of the three types of rocks: igneous, sedimentary, and metamorphic. Geologists find it helpful to use this classification, which takes into account the structure and the origin of the rocks. **Igneous** rocks have been formed by the cooling and hardening of molten material which has come from the interior of the earth, often by volcanic action. **Sedimentary** rocks have been formed by deposition by some transporting agent such as water, wind, or ice. **Metamorphic** rocks are igneous or sedimentary rocks which have been changed by having tremendous heat and/or pressure applied to them.

The Colorado mountains were formed by two major metamorphic uplifts. The mountains today are remnants of the youthful cycle of the second uplift, acted upon by volcanic, glacial, and erosionary processes. In ancient times there were several active volcanos in Colorado of much greater size than the present mountains. These volcanos eventually "blew their tops" and formed several smaller mountain peaks. The present shapes of the mountains are the result of erosionary action since the last great volcanic and glacial periods. The erosion continues; the mountain shapes are constantly changing. The mountains are dynamic.

But there is of course a stability and continuity to the land, and the following terms which help define the topography will be useful in your reading of this book.

A **peak** is a high mass, more or less conical in outline, that rises above its surroundings. Ordinarily a peak is a feature, or minor order, upon a range, but in some instances, as in the case of an isolated volcanic cone (the Spanish Peaks), a peak may stand alone and comprise the entire mountain mass.

A mountain **range** is an arrangement, usually linear, of many peaks, ridges, and their valleys. The term ordinarily applies to mountains that have general unity of form, structure, and geologic age. A couple of examples are the Front Range and the Sawatch Range.

A mountain **chain** consists of several associated ranges, usually more or less parallel, having unity of position, form, or structure, but separated by trenches or basins, such as the San Luis Valley, and Colorado's North, Middle, and South Parks.

A **cordillera** is a large regional grouping of mountain chains, such as the Rocky Mountains or the Andes.

A **massif** is a grouping of mountains forming a cluster of peaks and having unassociated ranges or mountains of irregular patterns and heights. Sierra Blanca is an example; it consists of several impressive peaks, all part of the same land bulk.

Between the mountain valleys there are uplands formed by remnants of the original elevation. These are called **divides**. The lowest and least massive features of mountain uplands are the foothills and spurs that fringe the principal uplands.

Foothills are hills located at the base of the higher mountains or hills, such as those at the eastern base of the Front Range.

Spurs are ridges projecting laterally from the main crest of a hill or mountain. A good example is the knife-sharp ridge of Capitol Peak.

A **pass** is any type of natural passageway through high, difficult terrain. The Colorado mountains have hundreds of natural passes, most of which were improved by man. They had such names as Mosquito Pass, Tincup Pass, and Whisky Pass.

A **col** is a depression in a summit line or a mountain chain, generally forming a pass.

A **cirque** is a natural amphitheatre or hollow excavation in a mountain system, formed by ancient glacial action or the erosion from an ancient volcano. A horseshoe cirque is an extended cirque that appears in the shape of a horseshoe.

A mountain **basin** is a large or small depression in the mountain form, such as the Pierre Basin on Capitol.

A **tarn** is a small mountain lake, often found in a basin, which generally has no apparent tributaries.

A **moraine** is the accumulation of debris carried down the mountain and deposited by glaciers. A lateral moraine is deposited at the side and a terminal moraine is deposited at the end of the glacier.

A **matterhorn** is generally a three- or four-sided, extremely sharp mountain or peak, quite pronounced in local relief, such as the alpine peak from which the term originates. Pyramid Peak is an example of this in Colorado.

The **steepness** of a mountain is usually less than it seems to an observer. The average slope of large mountains is seldom more than 20° to 35° from the horizontal; only a few have slopes of more than 35° near the summit. Even walls that seem vertical are seldom more than 70°.

The **Continental Divide** cuts an irregular north-south pattern through Colorado. It is formed by a continuous crest along the top of the mountains. Rain falling on the eastern side of the crest drains into the Gulf of Mexico. Rain falling on the western side of the crest drains into the Pacific Ocean.

Timberline is a pronounced line high on a mountain above which no trees grow. In Colorado, timberline is approximately 11,500 feet, although it is higher in some areas such as around Hoosier Pass. Its height is determined by the direction the mountainside faces, the prevailing wind, the protection afforded by the surrounding terrain, etc.

The geographic regions of Colorado run the gamut: the **Alpine** zone, above timberline, characterized by the lack of trees and the presence of only small wild flowers and plants; the **Sub-Alpine** zone, from timberline to about 10,000 feet, a zone of "black timber," including the Engelmann spruce, bristlecone pine, and balsam fir; the **Montane,** or "mountain zone," generally from 10,000 to 8,000 feet, where aspen, lodgepole pine, and Colorado's state flower, the columbine, abound; the **Foothills** zone, primarily along the eastern edge of the mountains, from 8,000 to 6,000 feet, characterized by sagebrush, yellow pine, Douglas fir, and mountain maple; the **Plains** zone, agricultural land, generally below 6,000 feet.

The Colorado Rockies were once considered a major deterrent to Western development; they were an obstacle to movement beyond. Then they became the object of westward movement; those Rockies were the destination for thousands upon thousands of adventurers seeking fortunes in gold and silver. The mountains have poured forth more metal and mineral riches than almost any other state, and are still doing so. Although mining income is now a smaller portion of the overall economy, the mountains remain the basis of Colorado prosperity through tourism. Colorado is called "The Switzerland of America." Many think of it as the ski capital of the world. The mountain slopes are wide and long, the snow is excellent for skiing, and the season is among the longest in the world. And the scenery is unsurpassed.

So today the Colorado Rockies are a major attraction for all America. We are still awed by them, but they have become our friends. We have given them names. And sometimes they have helped us find our own identity. We have learned that mountains are dynamic, that each has a history, a legend, a beauty, a life of its own. The following are just a few of many giant personalities found in Colorado.

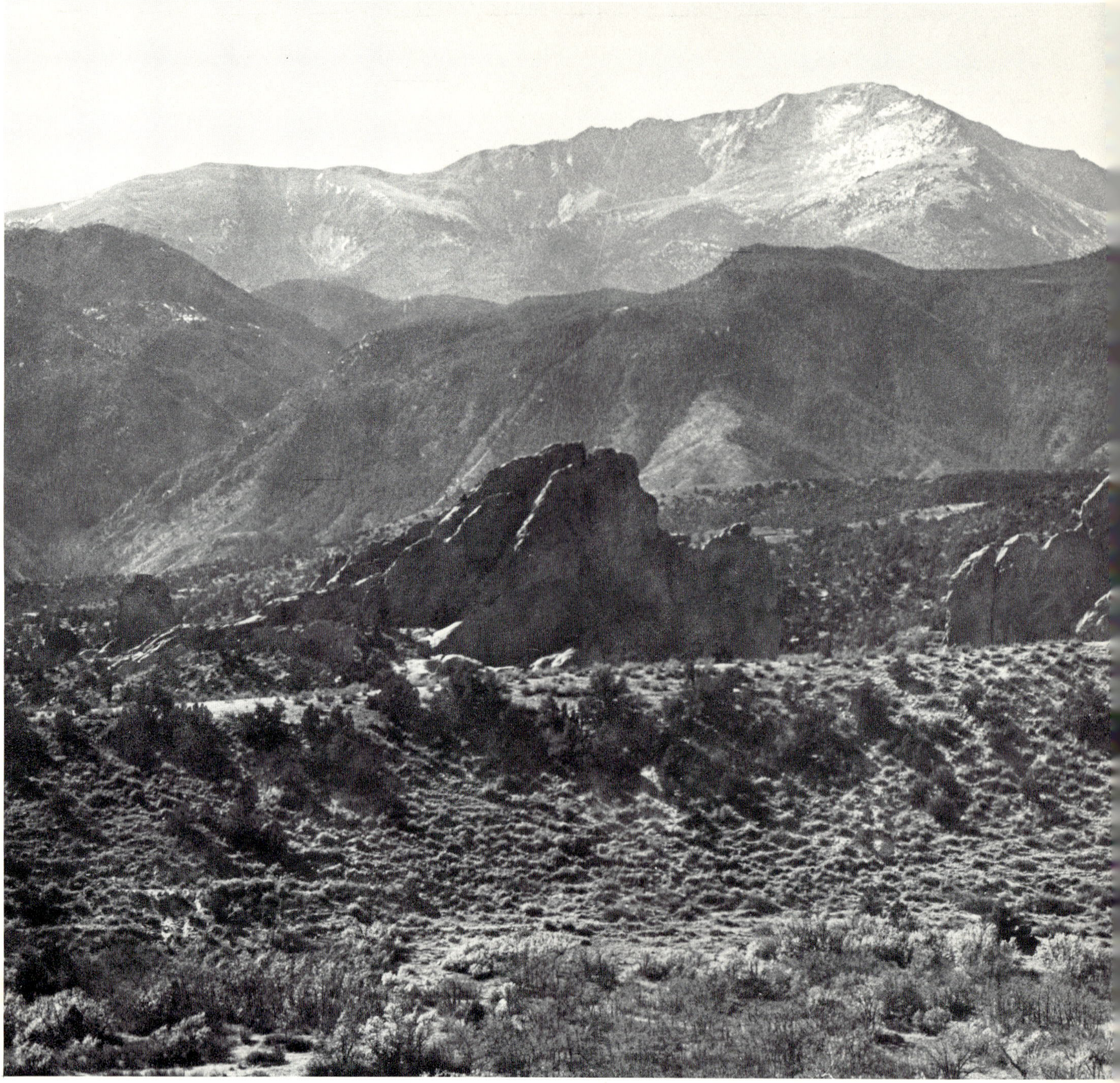

PIKES PEAK (Front Range) 14,110 feet
31st highest

Though there are many Colorado peaks which tower higher, Pikes Peak sums up for many Americans the whole of the state's mountain personality. The elevation and importance of the mountain were so bloated during the early years of the Colorado gold rush that many an easterner imagined that all of Colorado territory cooled in the shade of the mighty peak. Early mapmakers and guide books did little to diminish that belief. By 1858 and 1859, the rallying cry of the gold rush had become "Pikes Peak or Bust," for hopeful prospectors believed the peak was built on a hoard of endless treasure. However, the eager thousands who combed the peak found little or nothing, and the gold rush was soon labeled the "Pikes Peak Fiasco" or "Pikes Peak Hoax." It was not until thirty years later that the seemingly infinite Cripple Creek gold discoveries were made at the foot of the peak. Some historians believe the founding of the "greatest gold camp of them all" was actually delayed several years because of the disenchantment experienced by prospectors who took part in the first era of Pikes Peak gold fever. Other treasure has since been found near Manitou. The Pikes Peak region has also produced some of the finest smoky quartz ever found. Two crystal forms of smoky quartz from here are unique to the region.

The peak's riches were yet untapped when Lieutenant Zebulon M. Pike made his expedition in 1806 to explore the southwestern region of the Louisiana Territory. When he reached Pikes Peak, Zeb Pike estimated its elevation as 18,581, but his figure was off over 4,000 feet. Pike and a small party of his men tried to scale the mountain, but were able to reach only an intermediate summit. Though unsuccessful, this attempt made history as the first recorded try on any Colorado mountain. Because he and his men had to turn back, Pike predicted that the mountain would never be climbed.

The peak went through a series of names before one became generally accepted. There were many Indian names, the most common of which was "Long Mountain," for the mountain's long shape. Alexander von Humboldt's map of New Spain called it — or the whole conglomeration around it — "Sierra de Almagre." Pike referred to it as "Grand Peak," but mapped it merely as the wilderness' "highest peak." Major Long's expedition into Colorado in 1820 made short work of Pike's prediction that the peak would never be climbed and resulted in another new name for the mountain. Dr. Edwin James, botanist for the expedition, did the job easily and Major Long rewarded him by naming the peak "James Peak." Pike, however, was the first white man to discover and tell about the mountain, and most people regarded it as "his peak." It is believed that it was first called Pikes Peak by Colonel Henry Dodge in 1835. Expeditioner John C. Fremont called it Pikes Peak in the early 1840's, and by then that was the name by which the peak was known to most mountain men.

On July 6, 1858, Mrs. Julia H. Holmes became the first woman to climb Pikes Peak. With her was a party of prospectors from Kansas, who, a few days later, founded Montana City, a predecessor of Denver. It was nothing to climb the peak after that — and everyone did. In early Colorado City, it became fashionable to ride to the top of the mountain on a donkey. The climb is still popular and, despite heavy snow, is frequently made in winter. Annually, the Colorado Springs AdAmAn Club shoots off fireworks from the summit on New Year's Eve, a tradition that was begun in 1921 by Ed and Fred Morath. The following year the Morath brothers were joined by other climbers, and since then the club has added a new member each year.

In the early 1880's, the first attempt to build a cog railroad up Pikes Peak was abandoned, though a carriage road to the summit was completed in 1883. David Moffat and associates began the Pikes Peak Cog Railroad in 1889 and by June 30, 1890, the first train reached the mountaintop. The cog road covers 8.9 miles from station to summit, and rises at a 25° grade. Built at a cost of $1,250,000, this system remains the world's highest standard gauge railroad. The first auto ascent of the peak was attempted in 1900 by John Brisben Walker, colorful Coloradan and international figure. Though he reached only 11,000 feet, *The Rocky Mountain News* claimed that it was the highest point ever reached by auto in the world. The following year, a successful ascent was made in two days time by W. B. Felker and C. A. Yont, who reached the Pikes Peak summit in a Locomobile on August 23, 1901. Today, top racers in the country try out their daring and their cars in the annual Pikes Peak Hill Climb.

There are even those who run up the mountainside. What started out as a race up the peak between smokers and non-smokers in the mid-1950's has become an annual endurance race with top distance runners from throughout the state and the nation competing. The 20-mile course begins at the Cog Railroad Depot at Manitou, follows the old Barr Trail to the top, and back again. There are four divisions among competitors: Junior, Senior, Adult, and Old Timers. Some years, women have joined the race half the distance to the top.

For a brief time, a small tourist community existed halfway up the mountain on the cog railroad, and was appropriately named Halfway House Station. There was a two-story hotel, a restaurant noted for its fried chicken, and other tourist facilities. A town was laid out, several sites sold, and some cabins built before the settlement was found to interfere with the Colorado Springs water system. The city took over the property and the buildings were torn down in 1926.

Among those who came to Pikes Peak in the late 1800's, Sgt. John T. O'Keefe was one of the most colorful. In 1876, the U.S. Government established a meteorological station at the summit of the peak, and John O'Keefe was put in charge of it. It soon became apparent to the population of Colorado Springs and, in fact, of all Colorado, where O'Keefe's true talent lay. For the next three years, while he lived on the mountain, "The Pikes Peak Prevaricator," as O'Keefe came to be called, entertained the region with his dramatic stories. The most famous of these concerned the "Pikes Peak Rats." They were said to appear only at night and would gorge themselves on anything edible. According to O'Keefe, on one occasion an army of them attacked his family. He saved his wife by wrapping her in zinc plating, and, with stove pipes on his own legs, O'Keefe smashed scores of rats and electrocuted many more. However, before the rats retreated, they had eaten not only an entire quarter of beef, but the O'Keefes' two-month-old baby as well, leaving behind merely a peeled and mumbled skull. The Story and all its gory details was printed as true in **The Pueblo Chieftain, The Rocky Mountain News,** and other newspapers across the country. Even after the tale was proved a hoax, people long believed that rats infested the mountain. The curious came from great distances to see them and the survivors of the rat invasion. To satisfy these visitors, a group of men erected a mock gravestone on the mountain "... in memory of Erin O'Keefe, daughter of John and Nora O'Keefe, who was eaten by mountain rats in the year 1876."

"The Prevaricator" had stories of gigantic snow drifts, of being attacked by six mountain lions at once, and of his dramatic and ingenious escapes. Balaam, his donkey, and Seldom Fed, his dog, also became celebrated through O'Keefe's tales of their many narrow escapes from danger on the peak. Once O'Keefe invented a story that Pikes Peak was an active volcano. He reported finding a small open crater of molten lava, one little belch from which melted all the snow within half a mile. O'Keefe's fear was that one day Colorado Springs might share the fate of Pompeii and other great but ill-fated cities.

Explorers, frontiersmen, prospectors, settlers, and adventurers were drawn to Pikes Peak in the earlier days of this country's history. Incredible riches were mined from the mountain which Zeb Pike had once called unconquerable. With both the fictional and factual adventures passed down to us by the many men and women who were attracted to the region, Pikes Peak has become an American legend.

The present 19-mile drive up the mountain includes 16 switchbacks, and it is estimated that yearly about a quarter of a million persons drive to the top. Since 1948, when Colorado Springs took over the operations of the highway and the summit house, approximately five million visitors have journeyed up Pikes Peak. The road is kept open for skiers all winter as far as Glen Cove, the 11,525-foot mark. On a clear day especially, when one can see Denver, Pueblo, and almost into Kansas, the panoramic view from the summit of Pikes Peak appears to stretch without end.

LONGS PEAK (Front Range) 14,256 feet
15th highest

This is Colorado's — and perhaps the nation's — most interesting climbing mountain. More than 70,000 persons from throughout the world have already scampered up more than 20 routes to the summit, and enthusiasts continue to climb it at the rate of from 1,000 to 2,000 per year. Despite Longs Peak's popularity, only a handful of climbers have mastered some of the most difficult paths up East Face, which consists of about 2,000 feet of sheer granite. Microscopic ledges and pinpoint crags make up the few interruptions of the cliff wall, and they have been given satiric names such as "Broadway," "The Ramp," and "Bivouac." The most difficult of all Longs Peak climbs — straight up Diamond Face — was not successfully accomplished until 1960.

The peak was named for Major Stephen H. Long, who presumably discovered it during his expedition in 1820. Zebulon Pike, and others, however, certainly saw it before this. Early French trappers called it and Meeker Peak "Les Deux Oreilles" (The Two Ears), and the Indians called them "The Two Guides".

The Long expedition, which found Pikes Peak easy pickings, did not climb Longs Peak. In 1860, a fellow from Gold Hill, near Boulder, claimed he was the first man to climb the peak, but later experts believe he merely climbed Mt. Meeker. W. E. Andree and two friends claimed to have climbed the mountain in August of 1861, but Colorado Mountain Club expert Roger Toll believes they climbed Arapahoe Peak instead. William Byers, the first editor of *The Rocky Mountain News,* who did a little bit of everything, attempted to climb the peak in 1864. Upon reaching a point approximately 500 feet from the top, he decided that was about as close as any man could come to the summit. Jules Verne, the first great science fiction writer, had no doubts that the peak would be conquered, however. In his 1866 story, "From the Earth to the Moon," Verne envisioned the first projectile trip to the moon, fired from Florida. The progress of that imagined moon shot was observed from Longs Peak through a giant, 280-foot telescope with a 16-foot reflector which, in Verne's story, was mounted at the summit.

In 1868, four years after his first attempt, Byers was a member of the first official party to climb Longs Peak. With him were one-armed Major John Wesley Powell, his brother, Captain W. H. Powell, and four Illinois college students. On reaching the summit, however, the party came upon evidence of an Indian eagle trap. Tradition has it that the trap was used for many years by an Arapahoe Indian called "Old Man Gun" to get eagle feathers for his bonnet. It is said that he climbed to the top at night so the eagles wouldn't see him.

After the first official climb in 1868, the Peak remained in the headlines through the years as new routes were found and new records made. A member of the Byers-Powell party, L. W. Keplinger, was the first man to climb the East Face, in 1870. The first recorded moonlight climb was made in 1896 by H. C. Rogers. The first official night climb up the East Face of the peak was made in August 1931. The first climb of Zumie's Thumb on the East Face was made by three Colorado University students in 1951.

The most difficult climb of all was accomplished on August 3, 1960 by experienced California climbers Robert Kamps and David Rearick. The two men were the first to make the inch-by-inch climb straight up Diamond Face. It took them 52 hours to make the gruelling ascent, and Kamps and Rearick spent the night lashed to the sheer walls of the mountain. Others have since mastered Diamond Face.

Most of the climbing records were made and most of the trails were broken by three men who loved the mountain and devoted much or most of their lives to it: Reverend E. J. Lamb, his son, Carlyle, and Enos Mills, who was to Longs Peak what Thoreau was to Walden Pond.

The Lamb family came to Colorado in 1876 and established a permanent home at the foot of the peak two years later. The Reverend Lamb was an itinerant preacher who traveled great distances to preach in isolated communities. The rest of the time he climbed all over the mountain, breaking new trails and guiding hundreds to the top. He and his son cut the original "pony trails" and built the first shelters at the foot of the mountain. The Reverend, who habitually dressed in a heavy raincoat even in the dog days of summer, was almost killed when he slid on a 2,000-foot ice field. At the last minute Lamb was saved when he caught at a projecting rock. The ice field was called "Lamb's Slide" for many years, but is now known as Mills' Glacier.

Carlyle Lamb first climbed the peak in 1879 with his father, mother, and brother. The next year he became a professional guide. From 1880 until his retirement as a guide in 1902, he had led 146 trips and some 3,000 people up the mountain.

He sold his homestead in 1902 to a self-taught naturalist named Enos Mills who would make nature a popular subject and make Longs Peak one of the most famous mountains in the world. His Longs Peak Inn, at the foot of "his mountain," was a museum, a lecture hall, and a haven for nature lovers from throughout the world. The inn burned in 1906, but was rebuilt by Mills himself in a more rustic style. In his first 17 years as a guide, Mills led more than 250 trips up the mountain. With him as a leader, however, it was much more than just a climb. Mills is called the father of the "field trip," for his treks were fascinating nature tours known around the globe. He was said to be the first man to climb the peak in winter, in 1903; he broke many trails that were thought impossible. Among his several books about nature and his mountain, some are classics in their field; four of his most popular titles, still in print, are *Bird Memories of the Rockies, Forests and Trees, Watched by Wild Animals,* and *Wild Animal Homesteads.* Mills also led the fight to preserve the land surrounding Longs Peak and is called "the father of Rocky Mountain National Park." After his death in 1922, his wife and his brother, Joseph Mills, carried on his work.

Largely through the efforts of men like Mills and the Lambs, thousands of persons from every state and from throughout the world have climbed the peak. A five-year-old child has reached the top, as have men and women 70 and 80 years old. During a snowstorm in June, 1927, the first marriage took place on the summit, between Lucille Goodwin and Burl Stevens. In August 1880, half of a 12-piece brass band, including a tuba, made it to the top to play "The Star-Spangled Banner" and "Nearer My God to Thee." Foot races up and down the mountain have even been held. The record time of four hours and 17 minutes set in 1930 was not broken until 1946, and then by only six minutes. In 1939 the mountain was climbed twice by 20-year-old Dick Bice of Longmont — on crutches.

Most of the thousands of persons who have reached the summit have traveled the 15 or so conventional routes on the other sides of the mountain. However, about 1,500 persons have climbed the half dozen routes up the East Face. After Keplinger's climb of the East Face in the 1870's it was not officially climbed again until Mills did so in 1903. Professor J. W. Alexander of Princeton broke the first trail up the northeast face in 1922. Two years later he also made the first descent of the south ridge, and broke new trails up the southwest face and northwest ridge. Two European mountaineers broke new trails up the northeast and northwest faces in 1925. Keyhole Trail, now one of the more popular routes up the west face, was not discovered until 1940.

Since climbing Longs Peak became a popular outdoor sport, about one-fourth to one-fifth of the climbers have been women. The first woman to scale the peak was writer Anna Dickinson, in 1873. Shortly thereafter the mountain was mastered by Isabella Bird Bishop, leader of the international jet set in the 1870's and 1880's. She was guided up the lofty mountain by James "Rocky Mountain Jim" Nugent, one of early Colorado's more colorful characters. Cornelia Otis Skinner and Edna Ferber were two more of the famous women who made the climb. Elizabeth Burnell Smith set the women's record by allegedly climbing the peak 45 times. The first woman to have died on the peak was Carrie J. Welton, New York socialite and philanthropist, who perished during a climb in 1884. However, the most publicized tragedy was the death of Agnes Vaille, who attempted to be the first woman to climb the East Face in the dead of winter.

She died of exposure on January 23, 1925, and Herbert Sortland died attempting to rescue her.

Officially, since the 1860's, there are about 18 fatalities among climbers, both men and women; but some put the figure closer to 30. In fact, the newspaper story about the death of a commercial photographer in 1939 said he was the 30th victim of the mountain. A 19-year-old boy was killed in 1946, and there was another death in 1962. But for every tragic story of death on the mountain there are many more involving dramatic rescues.

Phenomenal for the number of climbers who attempt its slopes, and for the treacheries of some of its climbing routes, Longs Peak exhibits another distinguishing characteristic. On rare occasions, snow formations have appeared in the shape of a cross on the southeastern flank of the mountain. In 1901, one such cross appeared that was clearly visible from as far away as Denver.

The other peaks that make up the mountain are Mount Meeker (13,911) on the southeast ridge; Pagoda (13,491) on the southwest ridge; Storm Peak (13,335) on the northwest ridge; and Mount Lady Washington (13,269) on the northeast ridge.

MOUNT EVANS (Front Range) 14,264 feet
13th highest

Although Mount Evans is one of Colorado's most noted "tourist mountains," it is familiar to scientists as well. Experiments ranging from the effects of high altitude on rat reproduction to important tests in preparation for man's venture into space have been conducted here. The peak has been the scene of countless cosmic ray experiments, testing of the effects of nuclear explosions on the atmosphere, space and aviation experiments, meteorological observations and surveys, and biochemistry experiments. Many famous scientists from all over the world have studied or worked on Mount Evans, including Carl Anderson, R. A. Millikan, and A. H. Compton, three of the nation's Nobel prize winners in physics.

The University of Denver Cosmic Ray Research Laboratory was constructed at the summit in 1936. The building, containing two rooms, 20 by 24 feet, was built in Denver, cut into sections, and hauled up the peak by trucks. Over the years, other universities and organizations have also participated in developing the facilities. The walls of the laboratory can withstand a wind velocity of more than 150 miles per hour. Though it has been claimed that this was the highest such laboratory in the world, there are higher laboratories on Mount McKinley, and in Peru.

The road to the top is still believed to be the highest in the world. The first road, reaching as far as Echo Lake, was built in the early 1920's. A primitive road to the top was completed in the early 1930's, and an improved road was dedicated in 1939. "Flatlanders" and others not used to high mountain driving may argue that the road still leaves something to be desired. Yet the thousands of tourists who make the trip each year claim that having the wits scared out of them is part of the fun of driving to the top. Those who reach the summit are well rewarded, for one can view a distance of about 200 miles, and a total area six times the size of Switzerland. The $50,000 summit house was dedicated in 1941.

In 1863, famous artist Albert Bierstadt and fellow artist Fitz-hugh Ludlow were the first men to climb Mount Evans. A short time later Bierstadt began painting one of his Colorado classics, "Storm in the Rocky Mountains," using Mount Evans as a model. Ludlow also used the mountain as the subject of some of his paintings. Bierstadt named the mountain after his wife, Rosalie. In 1870, however, the territorial legislature rechristened it Mount Evans, in honor of Governor John Evans, Colorado's second Territorial Governor and a good friend of President Lincoln. A nearby mountain was named Mount Rosalie.

Like Pikes Peak, Mount Evans is a Colorado mountain well known in legend. One in particular was enough to send gold seekers swarming over the area for many years in a fruitless search. It was told that a young prospector was caught in one of the terrible snow storms for which Mount Evans is famous. Though he was able to make it back to a tie-cutter's cabin near the foot of the mountain, the struggle killed him. Nuggets found in the young man's pocket were assayed and found to be rich in gold. For years afterward, many treasure hunters combed the mountain seeking the nuggets' source, but without success. Another, more recent mystery occurred in November of 1941 when strange SOS mirror signals were seen to be flashing from the summit of the peak. The strongest telescopes in the area were turned on the peak, but the source of the signals was never found, and the occurrence remains a disturbing puzzle to this day.

Weather on the peak is extreme and unpredictable. From Denver, one can see the winds whip the snow near the peak and the clouds wrestle to bypass the mountain. In 1953, a busload of tourists was stranded for 12 hours near the summit by a midsummer snowstorm.

MOUNT BIERSTADT (Front Range) 14,060 feet
39th highest

This peak was named for Albert Bierstadt, the noted painter of massive landscape scenes. Bierstadt and the Rockies helped make each other famous. Bierstadt's "The Rocky Mountains," when exhibited in 1863, established his national reputation. Another well-known painting is his "Storm Over the Rockies," which depicts neighboring Mount Evans.

In the early days, Mount Bierstadt and the surrounding peaks were known as the Chicago Mountains. Bierstadt named the highest one for his wife, Rosalie; this was later called Mount Evans and the peak south of it was named Rosalie Peak (13,575 feet).

Apart from artists, prospectors and miners were among the first to climb Mount Bierstadt. Many early prospect holes can still be seen near the summit.

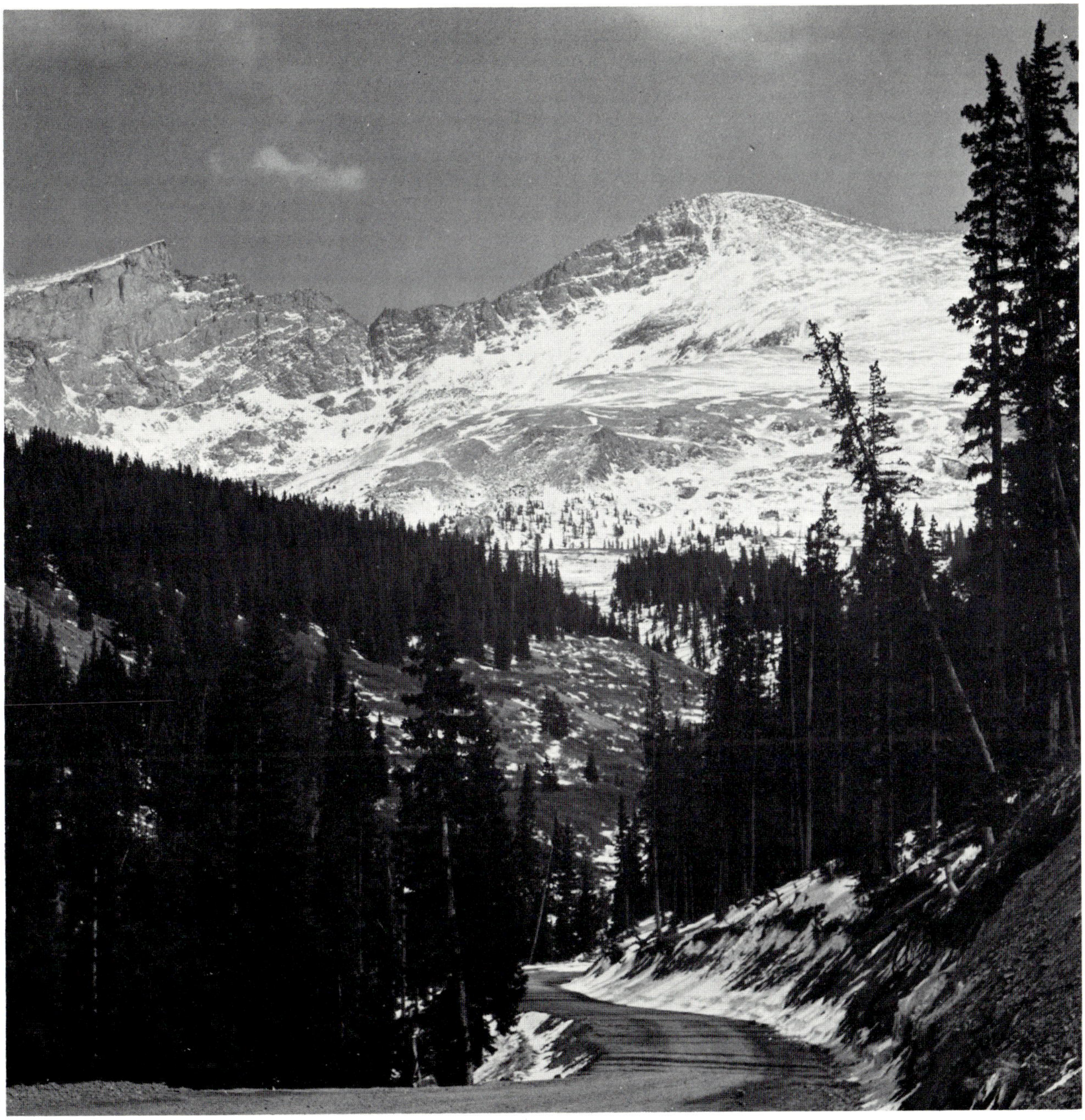

GRAYS PEAK (Front Range) 14,270 feet
9th highest

In talking about Grays Peak, it is difficult to separate it from its "twin," Torreys Peak. Because of their proximity to one another and their equal height (three feet difference), the mountains are often thought of as a pair. Both were popular early day climbs, when to climb Grays meant one was also planning to climb Torreys, and vice versa. (see Torreys Peak)

Asa Gray, for whom the peak is named, was a celebrated botanist and nature lover who was considered the fauna and flora authority of his day. Gray and botany professor John Torrey jointly authored many papers and books, the most famous of which was *Flora of North America,* published in 1841. Gray accompanied Charles C. Parry and other notables to the top of the peak in July of 1872 to officially christen this and the neighboring peak Grays and Torreys.

The popularity of early climbs up Grays and Torreys Peaks is demonstrated by the many detailed stories of climbs which appeared during the 1870's and 1880's in newspapers across the country and in national magazines such as *Scribner's Monthly.* Anybody that was somebody in Colorado made the climb, and the "challenge" attracted many tourists. A story of a night climb appeared in 1877. Pioneer photographer William H. Jackson helped popularize the mountain by making the climb himself; his photos of Grays Peak had wide circulation.

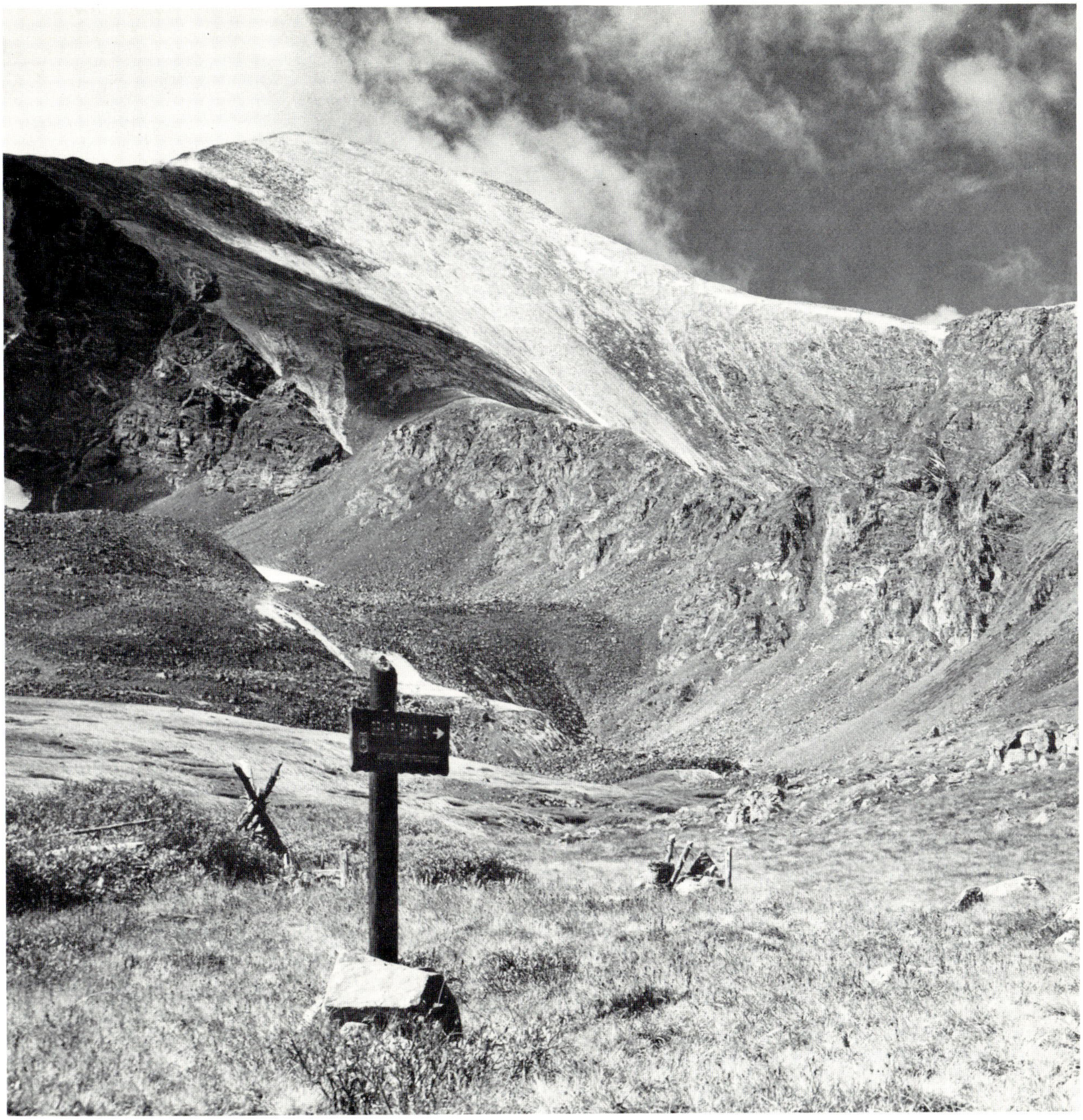

TORREYS PEAK (Front Range) 14,267 feet
11th highest

This and Grays Peak were almost as well known in early Colorado as Longs Peak and Mount Evans. During the boom years of Georgetown it was fashionable to climb the two peaks, and many excursions were made to the top. It was a "family climb;" groups and clubs made annual pilgrimages, and hoop-skirted ladies could walk or ride a horse to the summit. Torreys is separated from Grays by an immense cirque, which used to be the scene of many large outings. Both peaks were also known for many early nature study field trips popularized by John Torrey and Asa Gray.

Torreys and Grays Peaks had many early names, and some of them were interchangeable: Indians called them "The Ant Hills," and early prospectors referred to them simply as "Twin Peaks." Torreys was sometimes also called "Irwin Peak," after Dick Irwin, the pioneer mining man believed to have been one of the first to reach the top by horse. Before the names were made official, Grays was often called Torreys and vice versa.

John Torrey was a botany professor at Columbia University who compiled a classic book on Rocky Mountain flora, partly based on published accounts of plants noted by the 1820 Long Expedition into Colorado. Torrey gave names to many of the unusual wild flowers discovered in Colorado.

In the summer of 1872 the peaks were finally christened after Gray and Torrey during ceremonies at the summit, conducted by Charles C. Parry, a well-known scientist of the time. Some reports say that Torrey accompanied Gray and Parry and a party of other notables to the summit, but, in truth, Torrey was not in Colorado at the time. He did return to the peak a short time later, but his daughter made the climb for him because of Torrey's advanced age.

Although it is still a popular climb for families and beginners, Torreys no longer receives the publicity it once did. Generally, the peak is no challenge to alpinists, though some interesting and more exciting routes can be found. In July of 1950, two climbers fell during a climb and became trapped on a narrow ledge, where they were stranded all night. Their rescue the next day was one of the most dramatic in recent Colorado history.

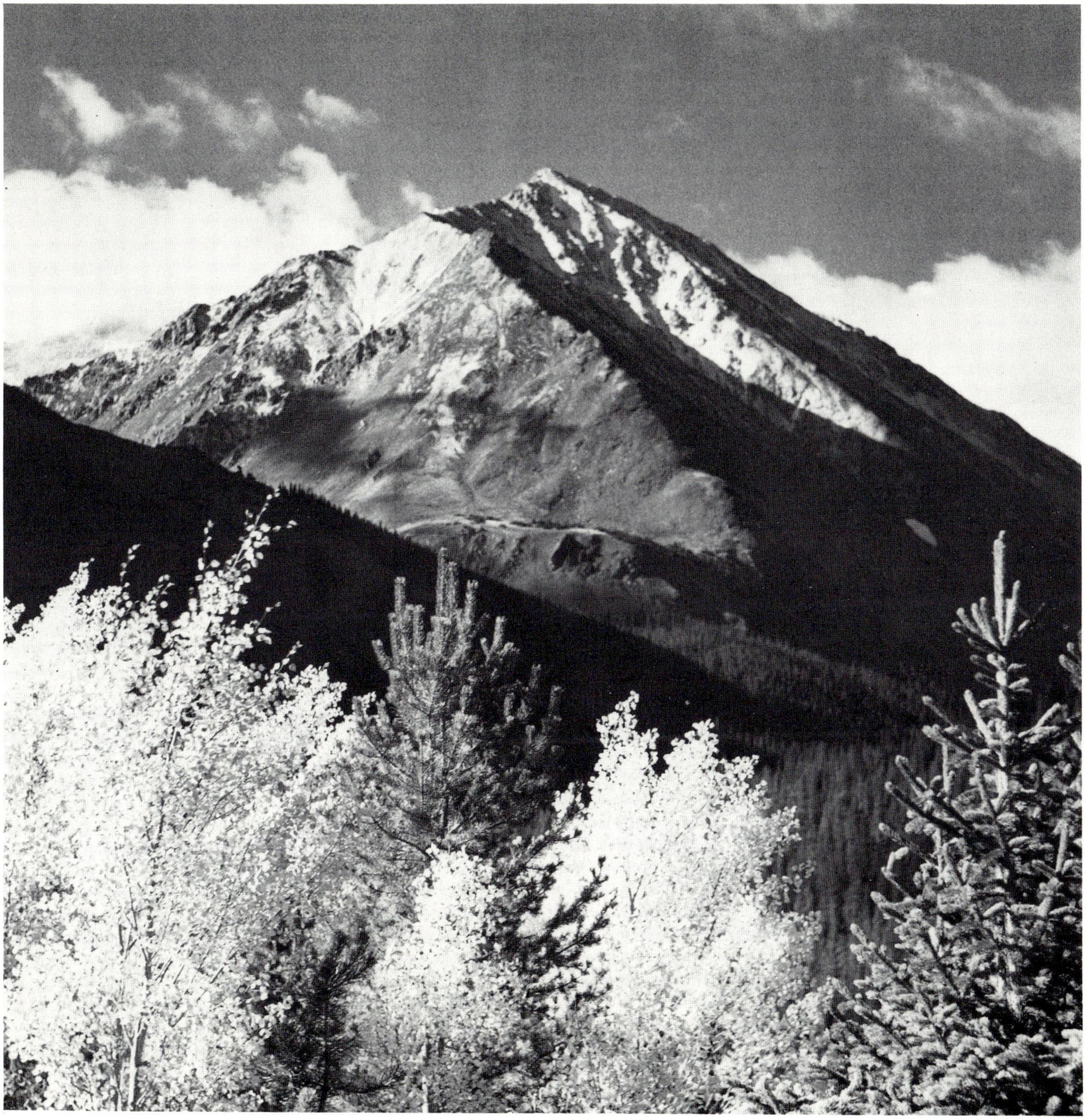

QUANDARY PEAK (Tenmile Range) 14,264 feet
14th highest

As is the case with many Colorado mountains, this peak bore various names before one was finally settled upon. The most common of the early names were Ute Peak, McCullough's Peak, and Hoosier Peak. The story of early miners who were in a "quandary" to identify some unusual rock found near the summit led to the name for both the peak and the most important mining town on the mountain, Quandry City.

Prospectors made their way all over the mountain, and much mining was done on all its slopes. Prospect holes can still be found near Quandary's summit.

Quandary is one of the peaks which has "grown" in recent years. The use of more accurate modern measuring devices has added 12 feet to its recorded height.

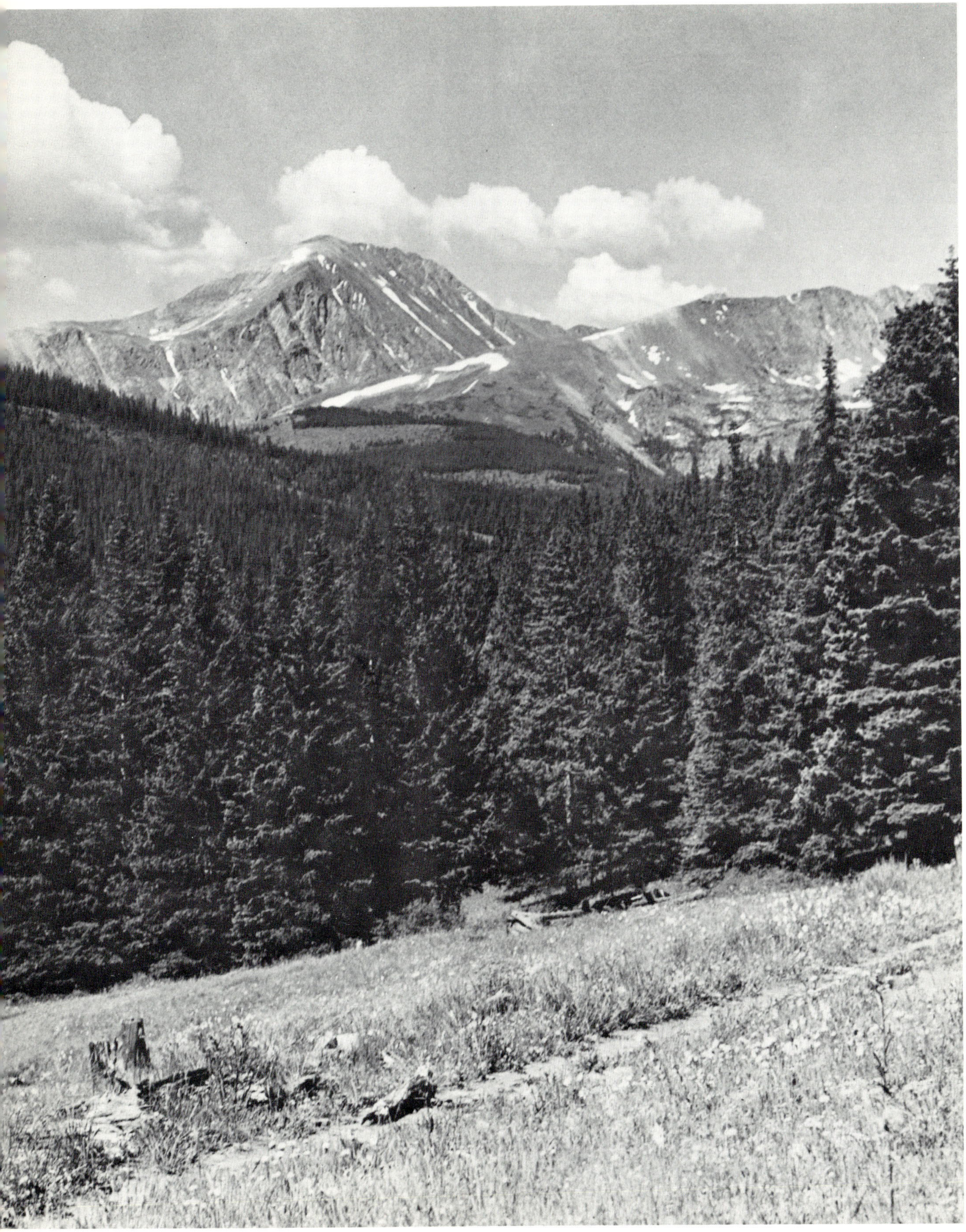

MOUNT LINCOLN (Mosquito Range) 14,286 feet
8th highest

Mount Lincoln has known other names and other altitudes. Some called it Mount Montgomery, for the mining town at its foot. Earlier, others called it Triaqua (Three Waters), because it was the source of three great rivers, the Platte, the Arkansas, and the Colorado (Grand).

At one time, this impressive mountain was believed to be the highest in the Rockies. In the mid-1860's, Bayard Taylor estimated its elevation at between 15,000-18,000 feet. About the same time, noted mining engineer, Professor Albert Dubois, estimated its height at 17,500 feet. One geologist said 17,300. Wilbur Stone, noted Colorado jurist and historian, organized the Mosquito Mining Company here in the early 1860's. In 1863, he climbed to the top of the mountain and estimated he was about 17,000 feet above sea level.

Exhilarated by his climb and greatly impressed by the magnificent mountain, Stone quickly returned to the mining camp near the foot of the mountain and called a meeting of the miners to decide on an appropriate name for the mountain. Many names were suggested before the name of the President was mentioned and such a unanimous chorus of agreement was raised that, as one story said, it "could be heard at the top of the mountain." The miners later sent the President a gold retort worth $800 which had been made from ore mined on "his mountain." Schuyler Colfax, personal friend of President Lincoln and a future U.S. vice president, journeyed to Colorado to personally thank the miners. His message from Lincoln was said to be the last one the President made before his assassination. It is also said that Colfax proposed to his future wife on the summit of Mount Lincoln.

Mount Lincoln was a "mining mountain" from the first. It is pockmarked with prospect holes and catacombed with mines. Millions of dollars in silver, gold, and other metals came from its innards. Miners worked through the winter in digging gold from the Russia, the Australia, and other mines high on its slopes, although some mines, such as the famed North Star, experienced problems with year-round frozen earth and ice. Early miners told of lakes of ice that never melted at the northeast foot of the mountain. A huge boulder rolled down the mountain in 1872 and smashed into one of the lakes, exposing a sheet of ice 12 feet thick, so it is said.

Surprisingly, a favorite sport of early Colorado miners was mountain climbing, and Mount Lincoln was one of their preferred peaks. It was not a "society mountain" like Grays or Torreys, and the miners loved it. They would even hold mountain-climbing races up and down the mountain.

It is still an interesting climb today, although not the most challenging. In addition to the many Fourteeners visible from its summit, some 50 peaks more than 13,000 feet high can be seen. Nearby Mount Cameron or Cameron Peak long basked in the prestige of being one of Colorado's highest mountains (14,238 feet), but the Geological Survey and Colorado Mountain Club have decided in recent years that it is part of Mount Lincoln and not truly a separate mountain at all. Only about 100 feet separate the top of Cameron and the ledge running to Mount Lincoln. So, instead of being a proud mountain unto itself, it is now called simply "Cameron Point" by the Colorado Mountain Club and relegated to being a mere "hump" of Mount Lincoln.

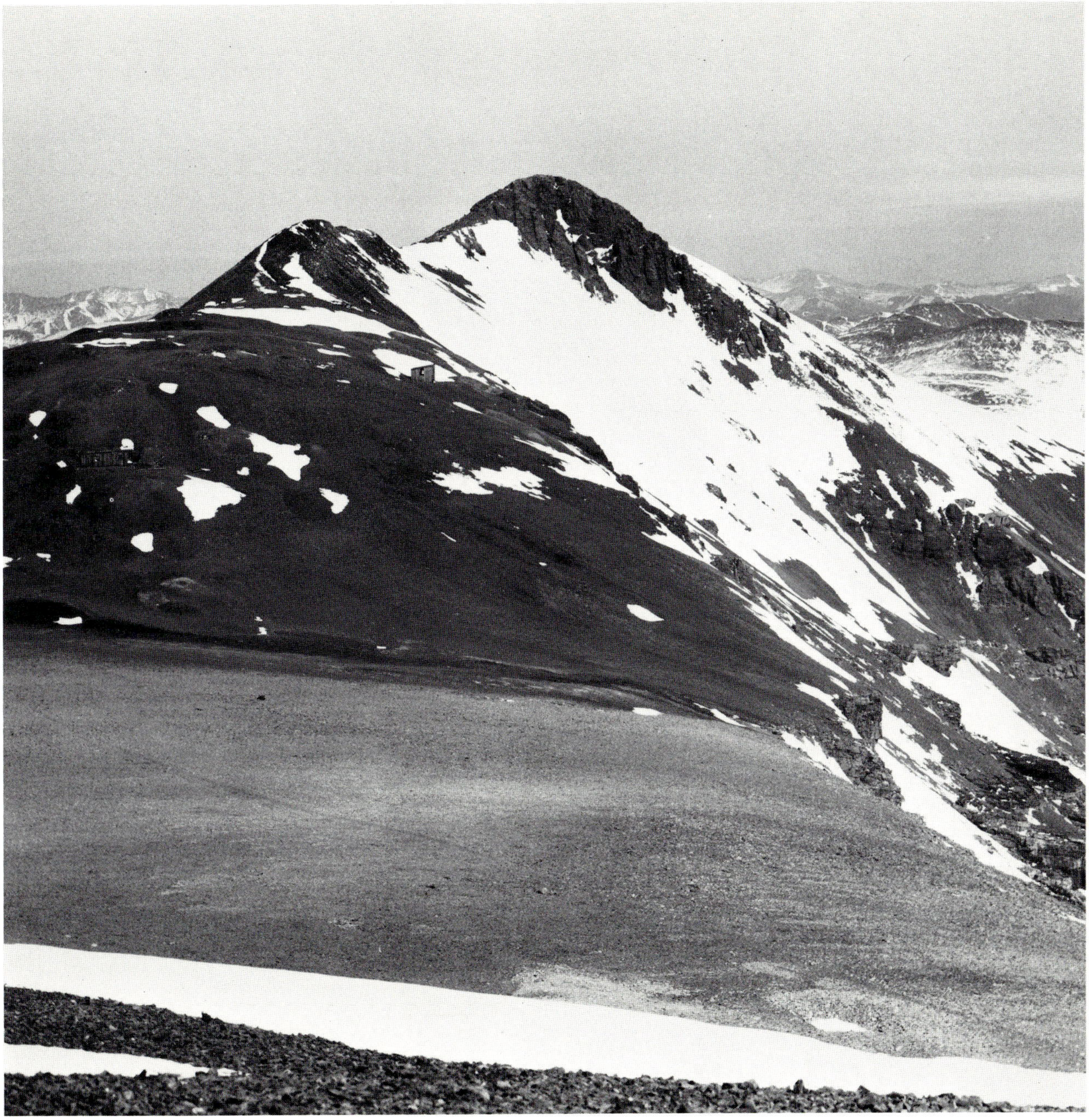

MOUNT DEMOCRAT (Mosquito Range)
14,148 feet
29th highest

Early reports tell that rebellious southerners during the Civil War had dubbed this peak "Mount Democrat," though U.S. Geologist Ferdinand V. Hayden later called it "Mount Buckskin" for the nearby mining town of Buckskin Joe. By the 1880's the rebel name had won out, for it was "Mount Democrat" that appeared on maps and official lists.

As the many prospect holes near the top testify, Democrat was an easy mark for miners and prospectors. They were able to climb all over it and many a claim was staked above 14,000 feet. The miners have long since gone, but many experienced climbers now "make" Mount Democrat, Mount Bross, and Mount Lincoln all in one day.

Perhaps the most remembered mining site on the mountain was the "Nigger's Cabin" located near the summit. It was occupied by a loner known only as "The Nigger," who mined a nearby claim for several years. Although the man kept to himself, it was known that he had amassed a considerable fortune from his solitary diggings. One day local people realized that he had gone, and never saw or heard from him again. Some suggested foul play, but most people believed the man had achieved the goal he had set for himself, then taken off to enjoy it.

After his sudden departure, hikers and miners used the well-built cabin until it burned to the ground in 1936. The blaze could be seen from as far away as Alicant Gulch in Leadville country, and miners there thought it was the glow of the planet Venus.

miners believed the mountain was sacred and that it was bad luck to deface it. Legend says one unbeliever began digging prospect holes in the side of the mountain and was soon crushed by a rock slide. Legend makers also claim that the figure formed near the cross is that of the Supplicating Virgin, and that the body of water near the foot of the peak is the "Bowl of Tears."

Mount of the Holy Cross was not made a Fourteener until 1964, and some wonder if it was so designated for its significance rather than for its actual height. During Colorado's 75th anniversary year, the mountain was further immortalized with a commemorative stamp. Yet, perhaps due to the crumbling cross and a diminishing faith in miracles or mountains, Congress erased Mount of the Holy Cross from the list of national shrines in 1954.

MOUNT MASSIVE (Sawatch Range) 14,421 feet
2nd highest

Mount Massive has been involved in more feuds than any other Colorado mountain, and the arguments have all centered around its height and its name. Some diehard oldtimers still claim that Mount Massive is higher than Mount Elbert. Massive does appear to rise above its neighbor, but Mount Elbert is actually 12 feet higher. There was also a controversy involving Mount Harvard, officially Colorado's second highest mountain, until new measurements recently reversed their standings. Some "Mount Massive people" thought that if their mountain could not be first, it should be second, and they even threatened to build a rock pile on top of the mountain to make up the two-foot difference between Massive and Harvard. But the U.S.G.S. got there first and gave a one-foot edge to Massive — without a rock pile.

Mount Massive won another, earlier height battle. A few years ago, Washingtonians were disapointed to discover that new official measurements showed Mount Rainier to be 11 feet shorter than Mount Massive. The Tacoma Chamber of Commerce threatened to build up their mountain to make it taller than either Massive or Mount Harvard in order than Mount Rainier would be the nation's third highest peak. That strategy, however, didn't work.

There were also several attempts to change Mount Massive's name. In 1922, an unsuccessful effort was made to change the name to Mount Gannett, for Henry Gannett, who was said to be the first man to climb the mountain, in 1874. Gannett was a leading topographer with the Hayden Surveys in Colorado and Wyoming during the period 1872-1879. In 1882 he became chief geographer for the U.S. Geological Survey and later was a founder and president of the National Geographic Society. He is usually referred to as "the father of American mapmaking." Again, in 1965, there was agitation to rename the mountain for Winston Churchill. This move was unsuccessful also, though not before it caused quite an uproar among the people of Leadville.

Mount Massive, the name apparently given it by early prospectors and miners because of the impressive appearance of the mountain, has remained. Will Rogers called this and Mount Elbert the "Ridgepole of the Continent."

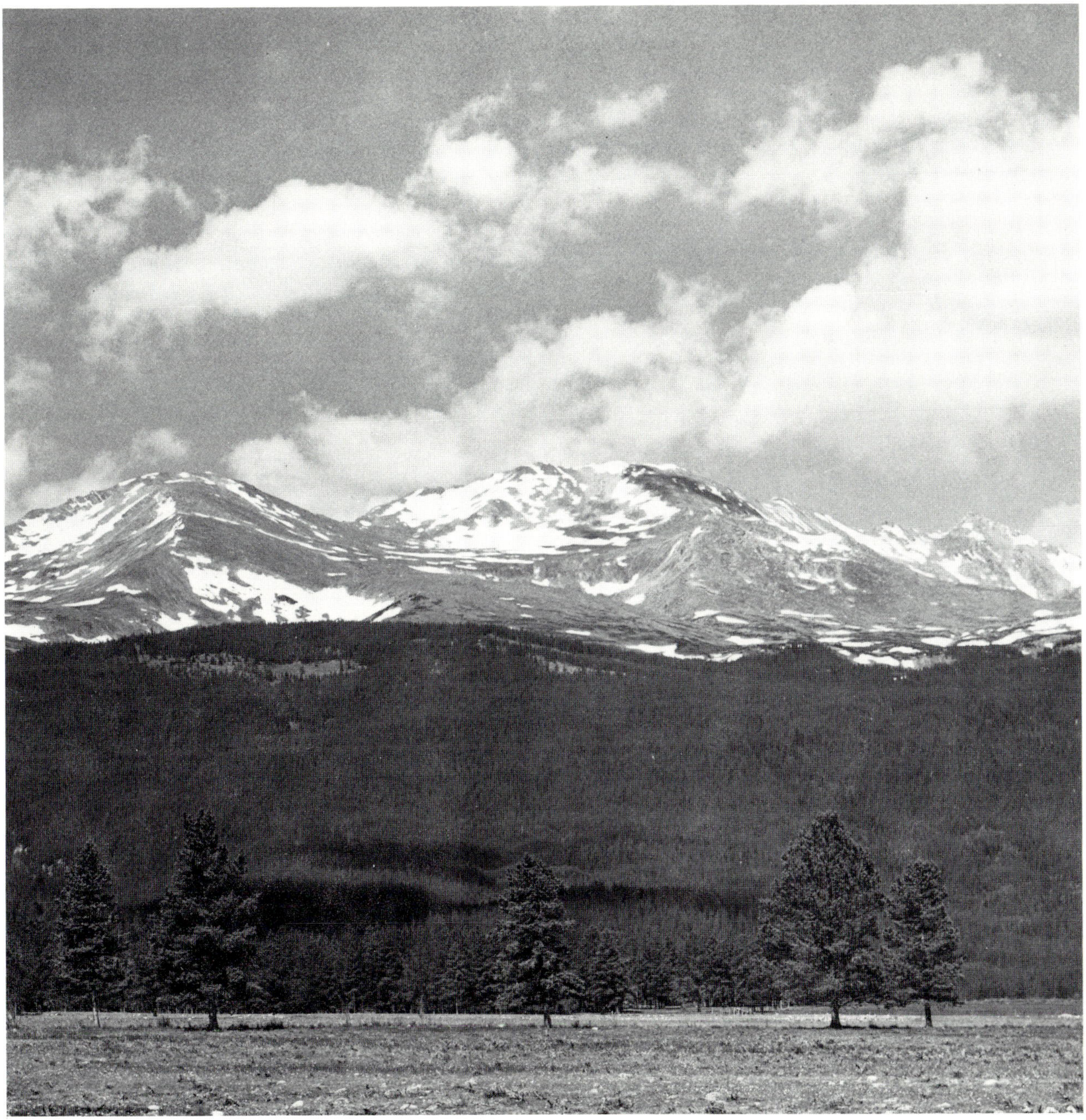

MOUNT ELBERT (Sawatch Range) 14,433 feet
Colorado's highest

Although its claim has been challenged in the past, Mount Elbert is now well-established as the state's highest and third highest in the United States. Despite its actual height, Mount Elbert is not too impressive in appearance, and many other Colorado peaks were once thought to be higher. In the mid-1930's, when Elbert was finally documented as Colorado's tallest, the "Mount Massive Fan Club" still didn't accept the Geological Survey's decision and they demanded a recount. When the demand was ignored, the club members threatened — as they also threatened about Mount Harvard — to build a rock pile on top of Massive that would overcome the few feet difference.

The mountain was named for Territorial Governor Samuel Elbert (1873-1874), who is credited with preventing Indian troubles in the San Juans and who married the daughter of Colorado's second Territorial Governor, John Evans. The mountain was apparently named before Elbert became Governor, however, since the name appeared on maps before 1873.

The first official climb up the mountain was made by H. W. Stuckle in 1874. It wasn't a noteworthy feat even at the time, since Elbert, for all its size, is considered an "easy walk-up" type mountain. In fact, every kind of vehicle has made it to the top. Early timber and mining roads took wagons to timberline and above as early as the 1870's. A jeep reached the top in 1949. Dave Morrison rode a 24-year-old bicycle to the summit in 1951. Taking a different approach, a *Denver Post* photographer, Albert Moldvay, delivered his paper to the top by helicopter in August of 1959. It may not have been the highest helicopter landing in the world, but it was the highest in Colorado.

Two Coot ATV's (all-terrain vehicle) made it to the top on August 3, 1969, the first vehicles to make it over the difficult and trackless northeast approach. The trip took about 24 hours and one of the three Coots that started didn't make it. Coot is a new ATV contraption that is not only four-wheel drive but also "four-wheel steer," with each wheel steerable separately. It can turn around within its own 7½-foot length.

Greatest of all the Fourteeners, Mount Elbert still has never had a road built to its top. The jeep that made it to the summit in 1949 was part of a larger caravan sent out to promote the construction of a road to the top and to boom a skiing area on Elbert. When one persistent jeep did succeed, broadcasts were beamed from the peak through lines and poles strung to Salida, while signal mirrors also flashed the news to Leadville. Promoters believed that since lesser Colorado peaks have summit roads, the state's highest peak should be accorded the same distinction. No modern road has yet been constructed, although one can drive to timberline over old roads. Similarly, despite Elbert's excellent potential, no major skiing development has occurred on the mountain. Nevertheless, it is still a favorite haunt for the "old-fashioned skier" — the cross-country and climbing skier.

The dynamism or ever-changing face of mountains is dramatically illustrated on Mount Elbert. In the summer of 1937, a violent storm shattered the steep rock face on the mountain's northwest slope, creating a new sharp cliff several hundred feet long.

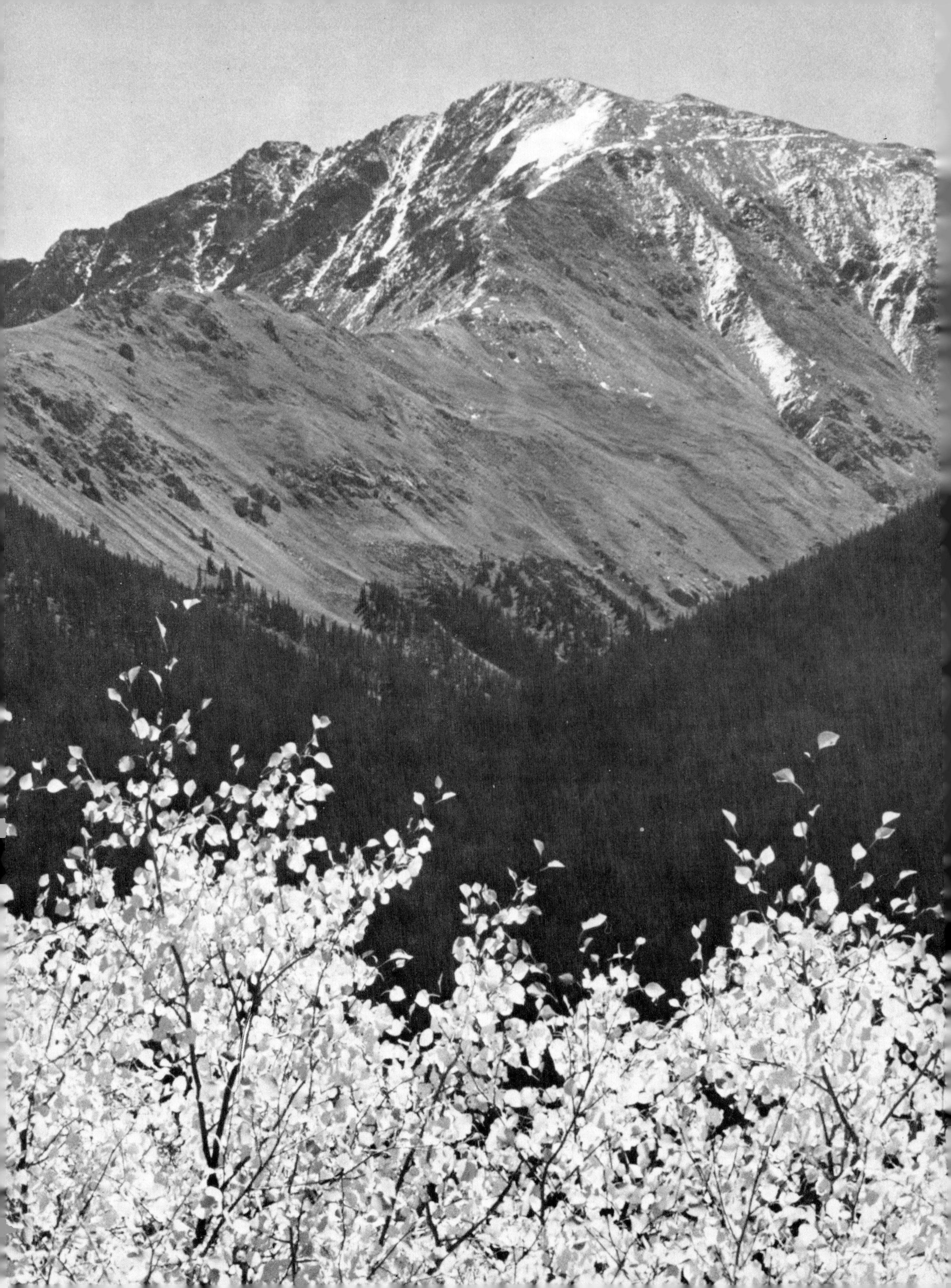

LA PLATA PEAK (Sawatch Range) 14,336 feet
5th highest

La Plata is separated from Mt. Elbert, Colorado's highest mountain, by Independence Pass, and most travelers are more impressed by La Plata than Elbert.

Ferdinand V. Hayden, the U.S. Geologist who named so many of Colorado's peaks, also chose the name for this mountain. "La plata" is Spanish for silver, and Hayden must have been thinking of the immense mining potential here. Prospectors were in fact the first palefaces to climb the mountain, and much evidence of their activity can be found close to the summit. They were convinced that the best lodes were usually found above timberline and often said:

> A good silver mine
> Is above timberline
> Ten times out of nine.

A route which lies along the rugged northeast ridge of La Plata is one of the few challenging climbs in the whole Sawatch Range. Ellingwood Ridge, as it is called, was named for Albert R. Ellingwood, who was believed to be the first man to negotiate that route, in 1922.

MOUNT OXFORD
(Sawatch/Collegiate Range) 14,153 feet
27th highest

This was a forgotten mountain for more than 50 years. It was first surveyed in 1873 by Hayden; he noted it and two other mountains in this range, all three over 14,000 feet, but he did not give them names. He only listed Mount Harvard and La Plata Peak.

Oxford was "rediscovered" about 1925, when it was duly noted and welcomed into the Fourteener family. A short time later, the Colorado Mountain Club named it, in keeping with the other names in the Collegiate Range.

A kind of snubbing of Mount Oxford continued. For years it was thought to be barely a Fourteener and was accordingly ranked 51st highest. More recent measurements awarded the peak about 150 additional feet.

Despite the long neglect, Mr. Oxford is today a popular "family climb" with its gentle slopes and broad, beautiful vistas. Perhaps the most popular starting point is from the old ghost town of Vicksburg.

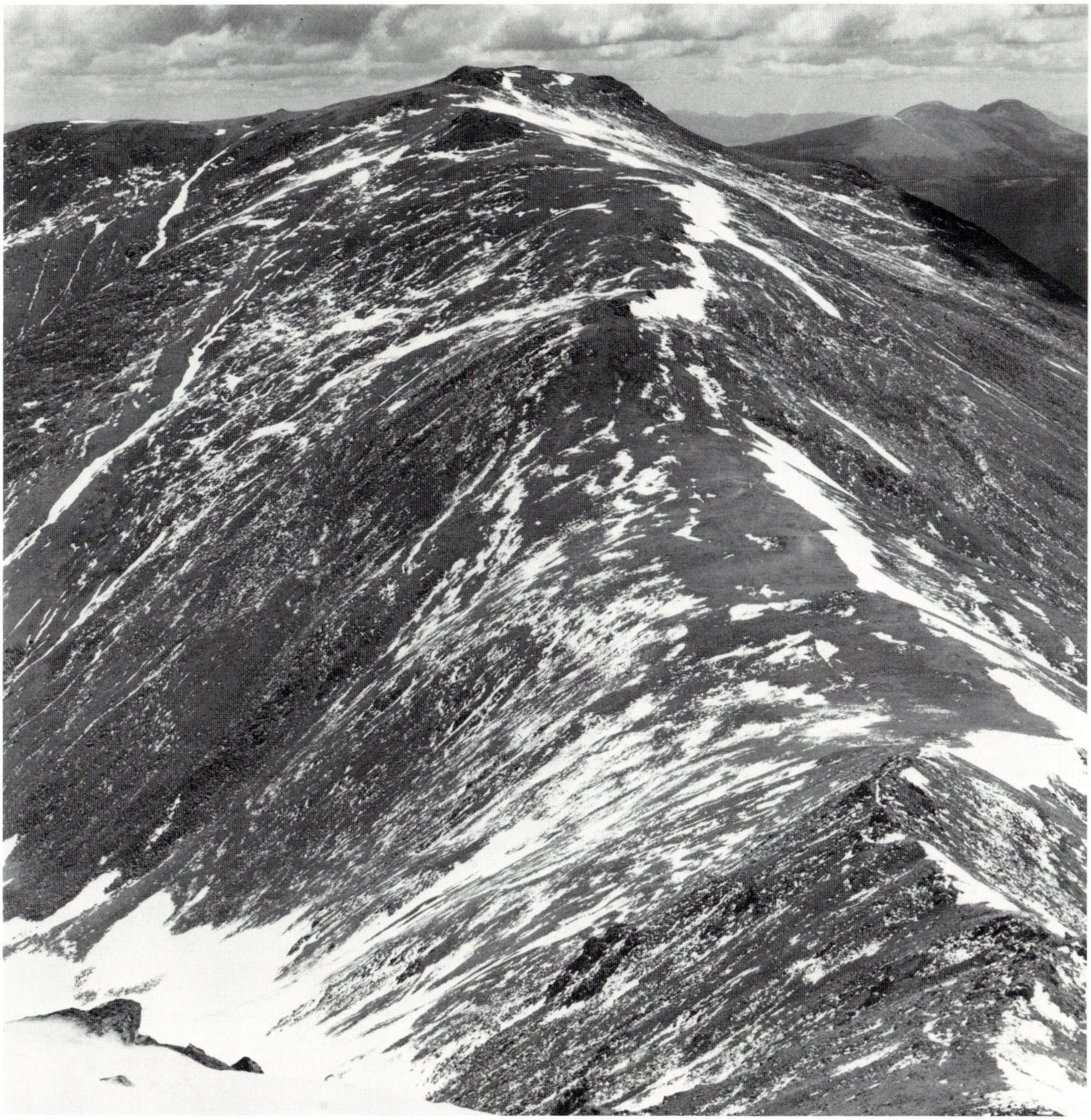

MOUNT BELFORD (Sawatch Range) 14,197 feet
19th highest

Although there is no official record, it is believed that this mountain was named for Judge James Belford, "The Red-Headed Rooster of the Rockies." Belford was appointed to Colorado's Territorial Supreme Court by President Grant in 1870, and remained on the Court until 1875. A year after his retirement from the bench, he became Colorado's first elected U.S. representative. Early in his Congressional career he was dubbed "The Red-Headed Rooster from the Rockies" because of his fiery red hair, matted beard, and flamboyant manner. He was known also as a strong drinker and as an orator of great wit. At any time, Congressman Belford might straggle into the U.S. House of Representatives, make a loud and colorful speech that may or may not have been pertinent to the subject at hand, and saunter out again.

He finally returned to Denver to practice law and write until his death in 1910. Mrs. Belford was a much beloved member of Denver society, known for her many charitable activities. A son, Herbert Belford, was a prominent Denver and New York City newspaperman and lawyer.

Ironically, Mount Belford is a rather colorless hunk of mountain, long neglected by both record-keepers and historians. Until recent years it remained unnamed and unmeasured since it was considered to be one of the three peaks belonging to Mount Oxford. Recent surveys have added about 140 feet to its official elevation and more prestige to its position among Colorado's Fourteeners.

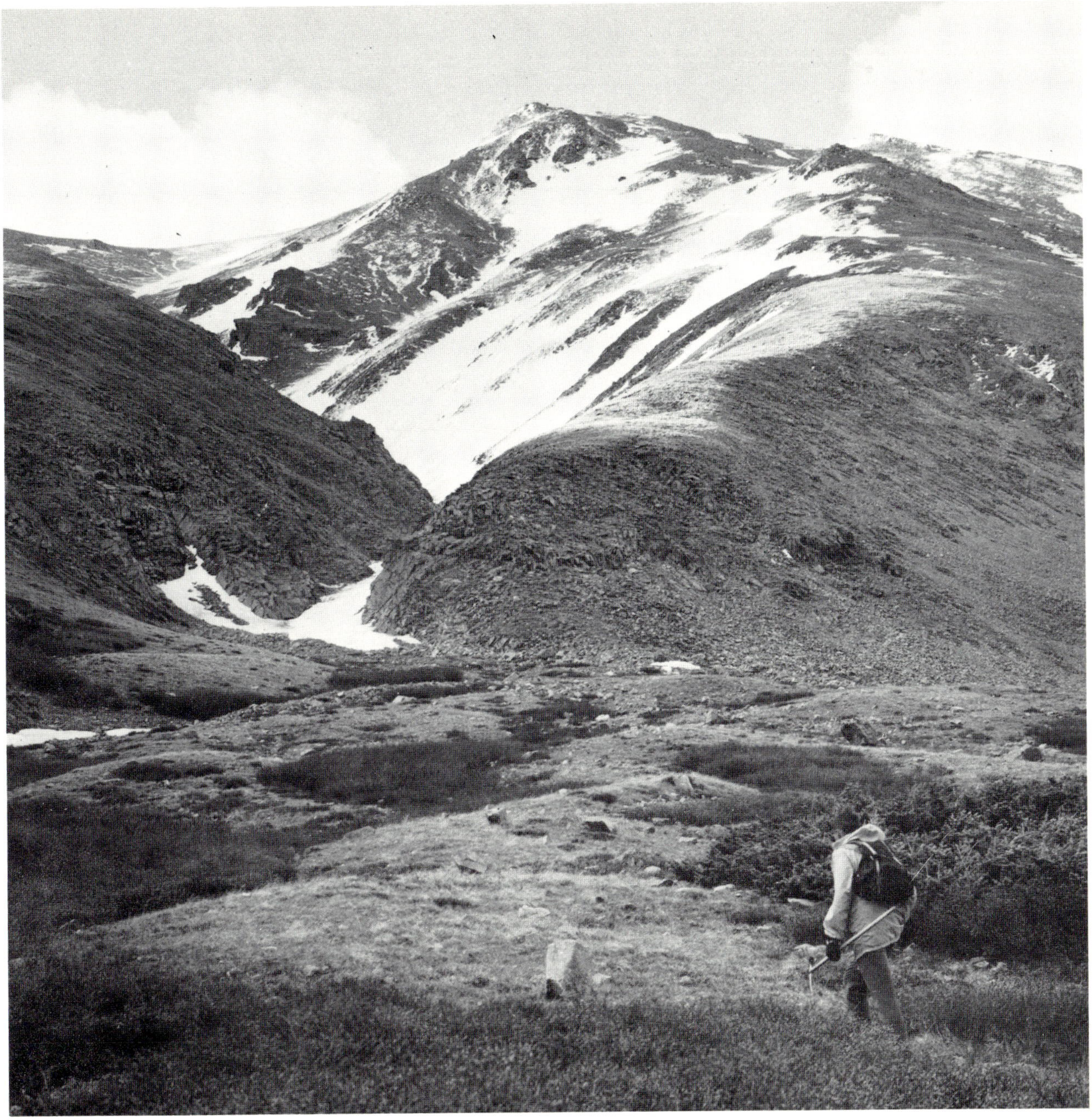

HURON PEAK (Sawatch Range), 14,005 feet
51st highest

There are several important similarities between Huron Peak and Mount of the Holy Cross. Both are the same height and both were late in being admitted to the family. Along with Missouri Mountain, Huron was admitted in 1956, only a few years before Holy Cross was included.

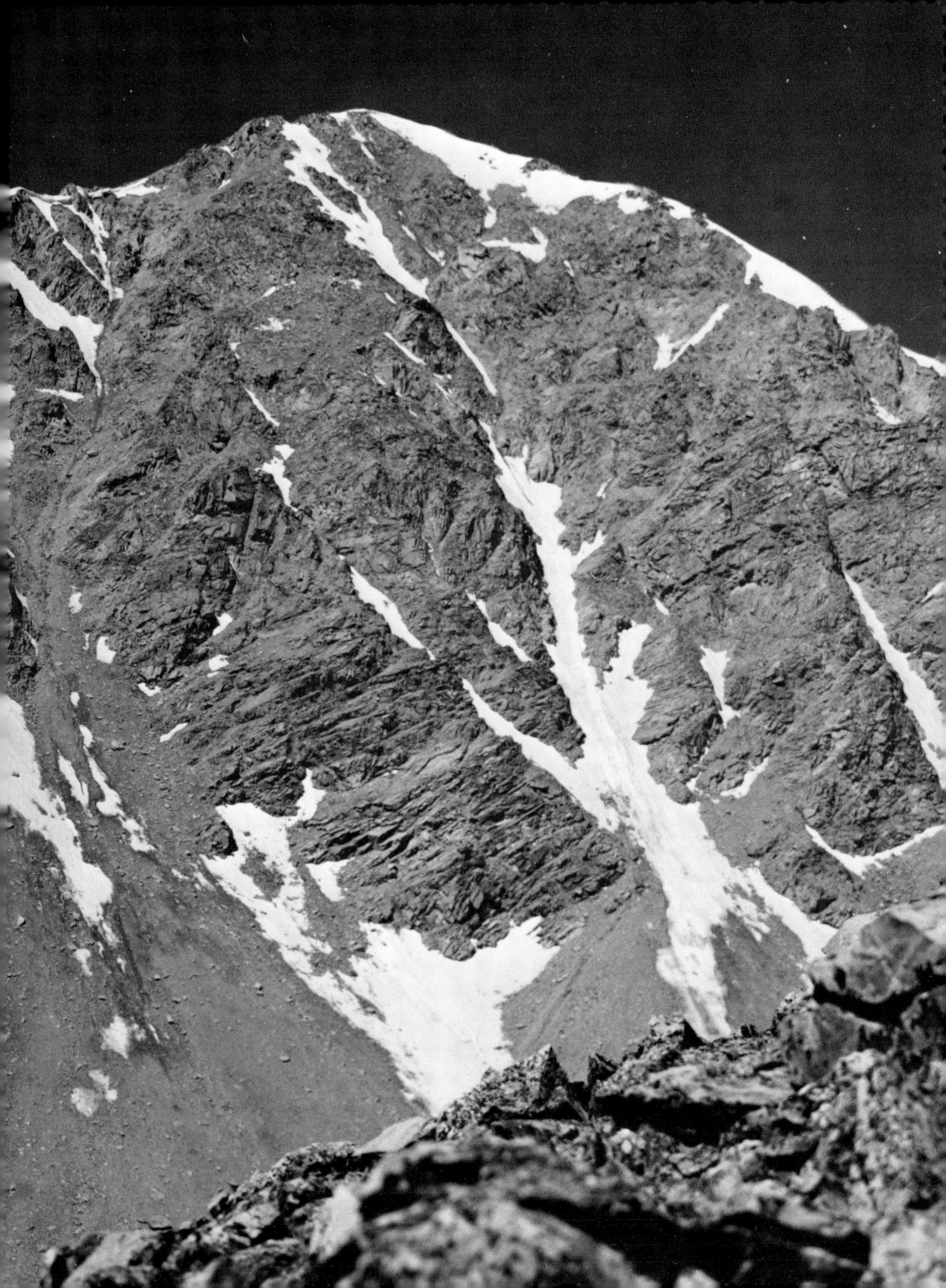

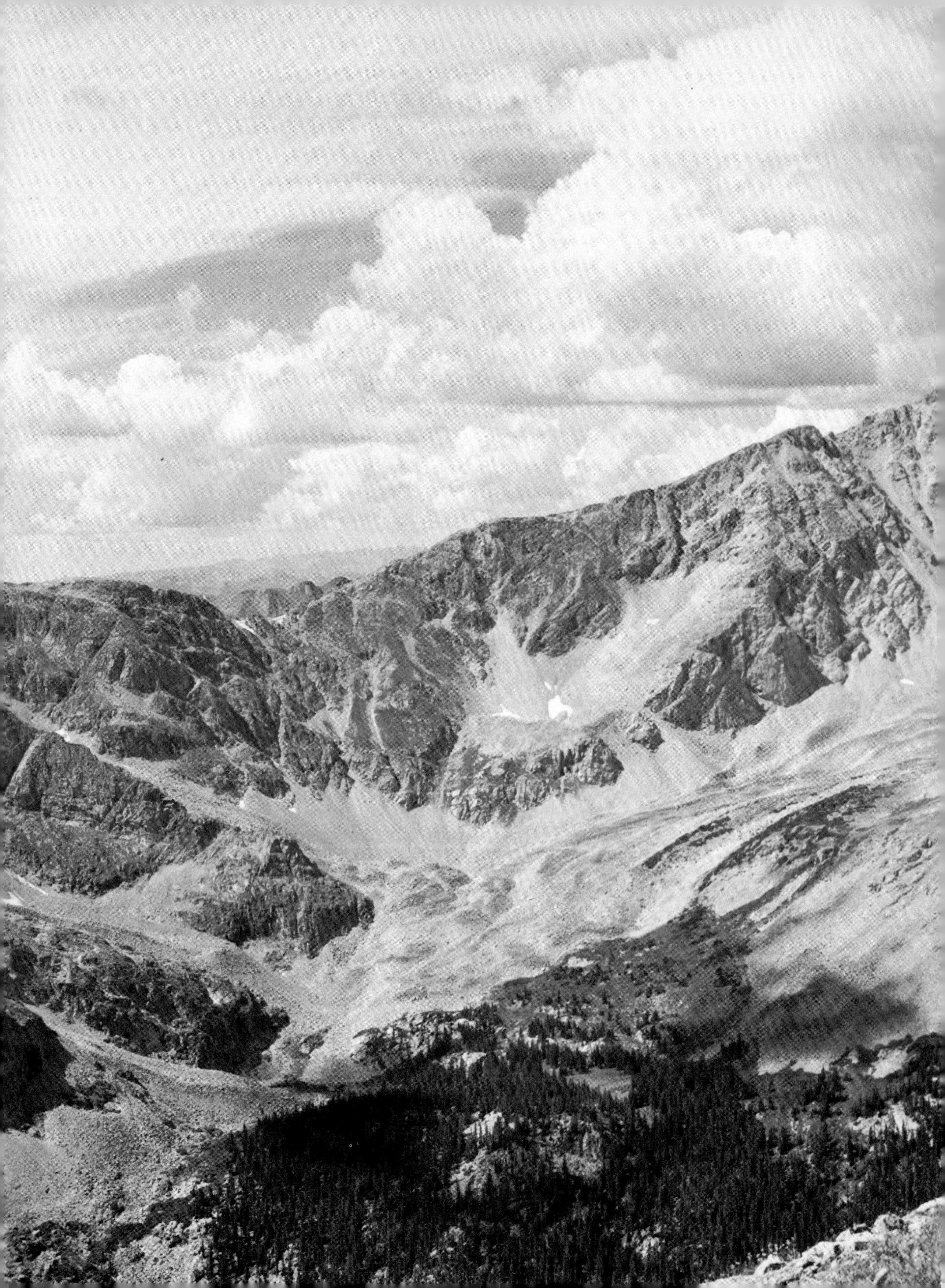

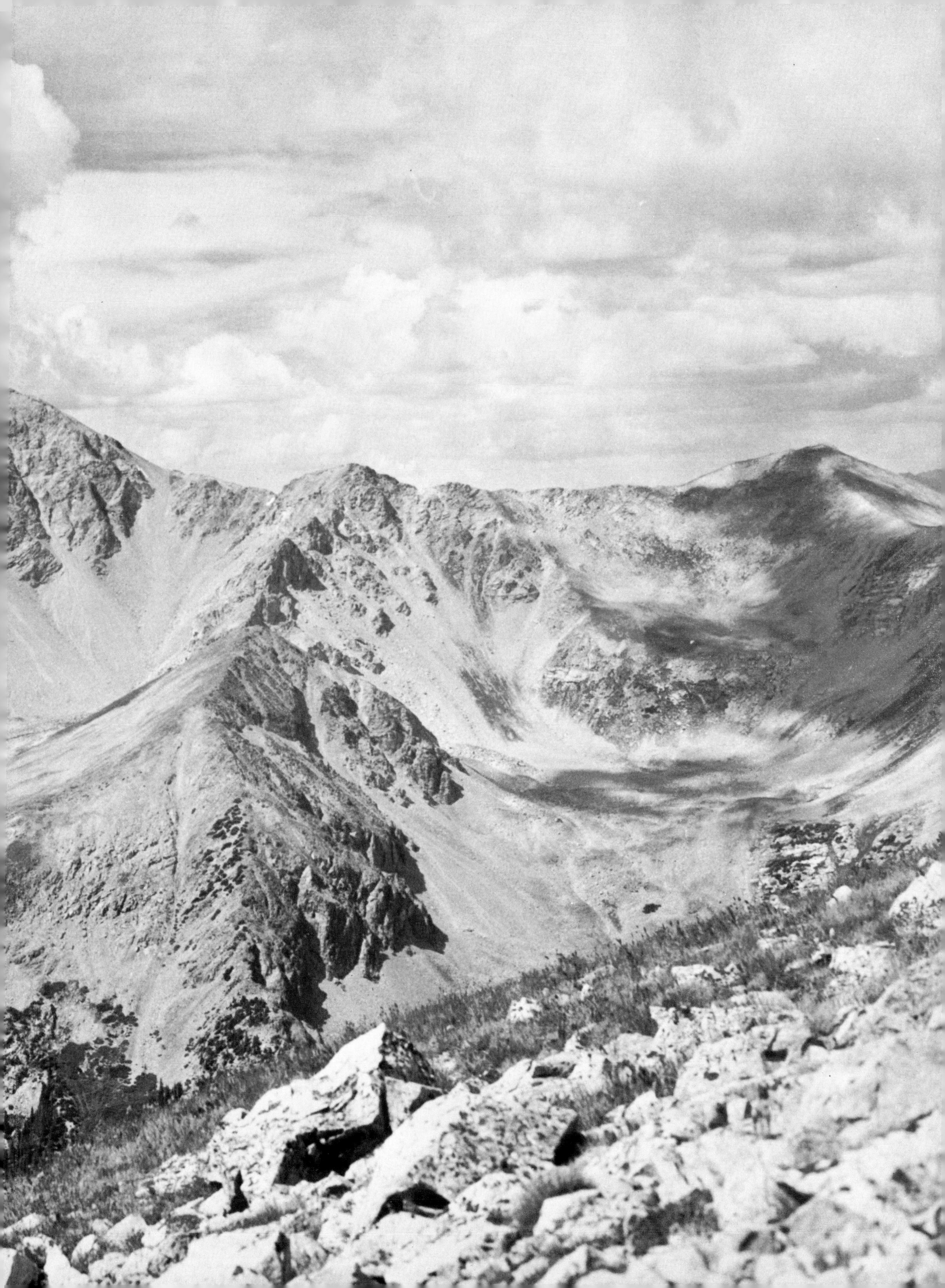

MISSOURI MOUNTAIN
(Sawatch Range) 14,067 feet
36th highest

Until recent years, this was one of three peaks considered a part of Mount Oxford. It was given a separate identity and a specific elevation by the Geological Survey in 1956. It is connected by ridges to Mount Oxford and Mount Harvard, yet the U.S. Geological Survey calls it a separate mountain.

Most commonly known as Missouri Mountain, it is sometimes refered to as Missouri Peak.

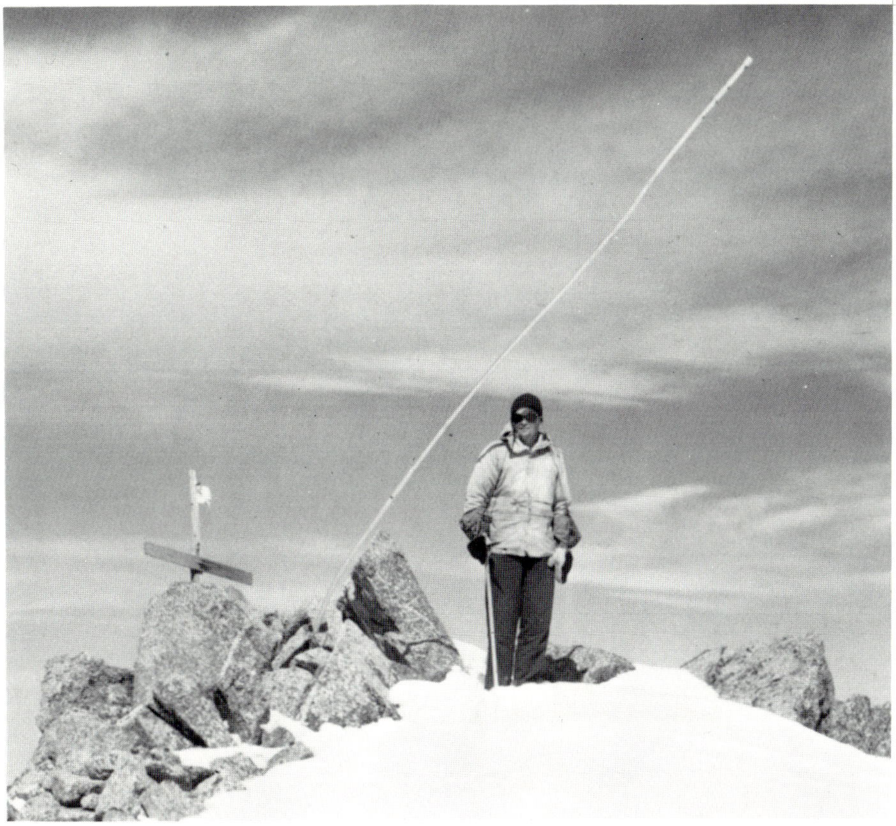

MOUNT HARVARD
(Sawatch/Collegiate Range) **14,420 feet**
3rd highest

The mountain was named by Professor Josiah Dwight Whitney, who surveyed much of Colorado and for whom Mt. Whitney is named. He brought Harvard's first mining school graduates here in 1869. The group climbed the peak, believed it to be the highest in Colorado, and named it for their school. This was also the beginning of the naming of the Collegiate Range.

A few years ago some Harvard students attempted to erect a 20-foot aluminum pole atop the mountain in order to make it higher than any other Colorado peak. Apparently they were not as bright as most Harvard students, since they left the pole on a false summit. The job was correctly done a couple of years later by some Cornell men, who then wanted to call the mountain Mount Cornell.

Despite the difficulty encountered by the Harvard students, Mount Harvard is an easy and satisfying climb. Generally, one can jeep up to the famed old Leonhardi Mine. Other mines and prospect holes indicate prospectors were able to maneuver all over the mountain.

In the winter of 1956-57, the eyes of the animal-loving world were focused on a 12,500-foot saddle between Mounts Harvard and Yale where an old bay horse was spending the winter. Airline pilots, who sometimes went out of their way to show the horse to passengers, named the horse after Elijah, the biblical prophet who was fed by ravens in the wilderness. People around the world sent money to help finance "Operation Haylift," through which planes dropped tons and tons of hay to the horse. Running accounts of Elijah's welfare and later rescue appeared in newspapers as far away as London and Paris.

The animal turned out to be a well-known local horse named "Bugs," who had led many a saddle jaunt out of Buena Vista. His name had been given him because of his uncommon aversion to automobiles and women — especially women in slacks. The owners believed this was why Bugs had gone on up the mountain — "to get away from it all."

MOUNT COLUMBIA
(Sawatch/Collegiate Range) 14,073 feet
35th highest

Along with Mount Oxford, this was one of the last peaks in the Collegiate Range to be given a name. It was first named by Roger Toll of the Colorado Mountain Club in 1916 when he climbed the peak; the name was officially adopted by the club in 1922.

The peak is southeast of Mount Harvard and connected to that mountain by a jagged ridge which offers a rugged challenge to mountain climbers.

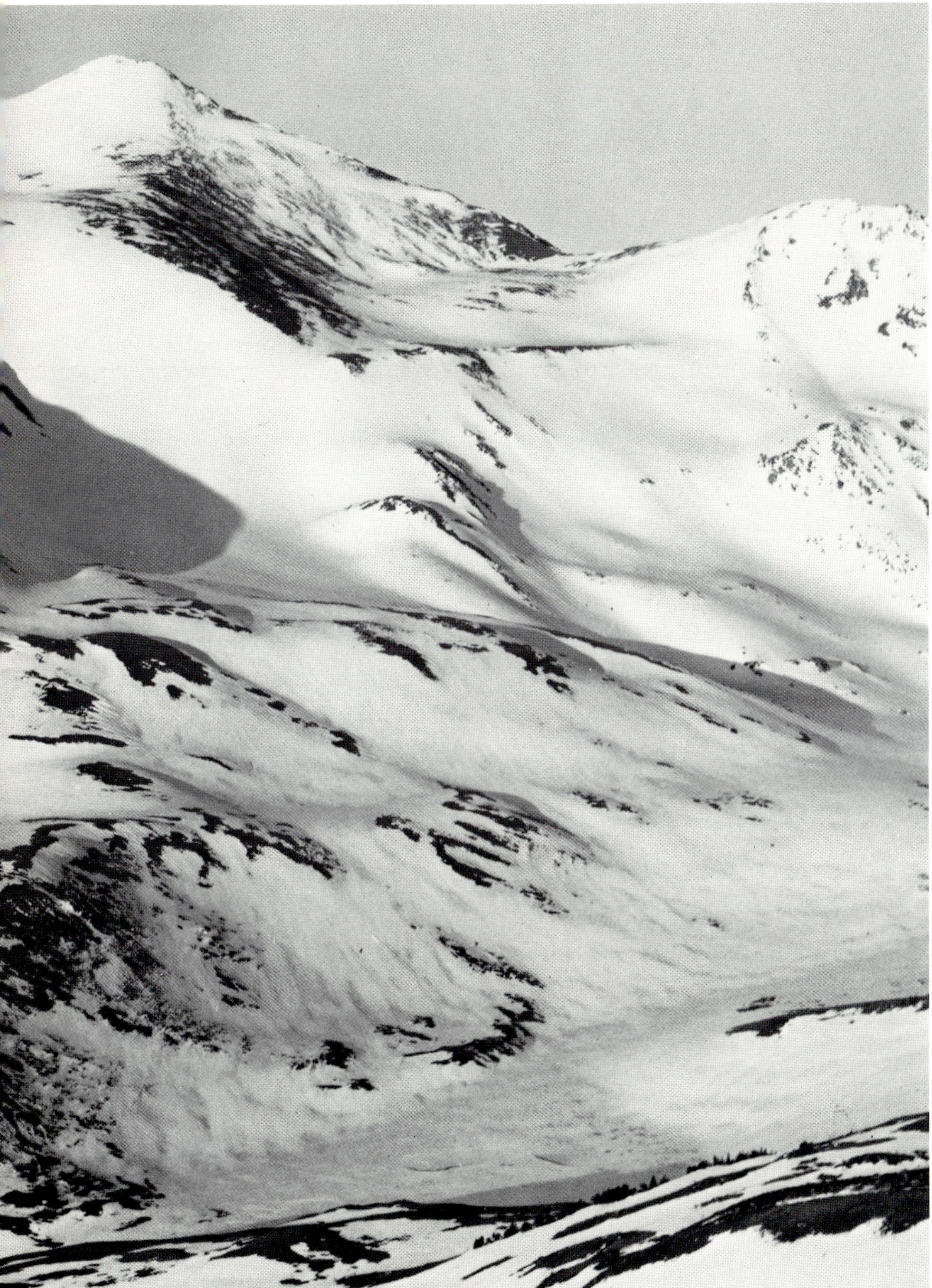

MOUNT YALE
(Sawatch/Collegiate Range) 14,196 feet
20th highest

This was another peak named in 1869 by Professor J. D. Whitney when he was in Colorado with his Harvard Mining and Mountain School students. Although a Harvard professor and founder of the mining school at Harvard, Whitney still had loyalty to his alma mater, Yale.

This is the third highest peak in the Collegiate Range and offers little challenge to mountain climbers. It was believed first climbed by prospectors, and there is still some mining done on the mountain.

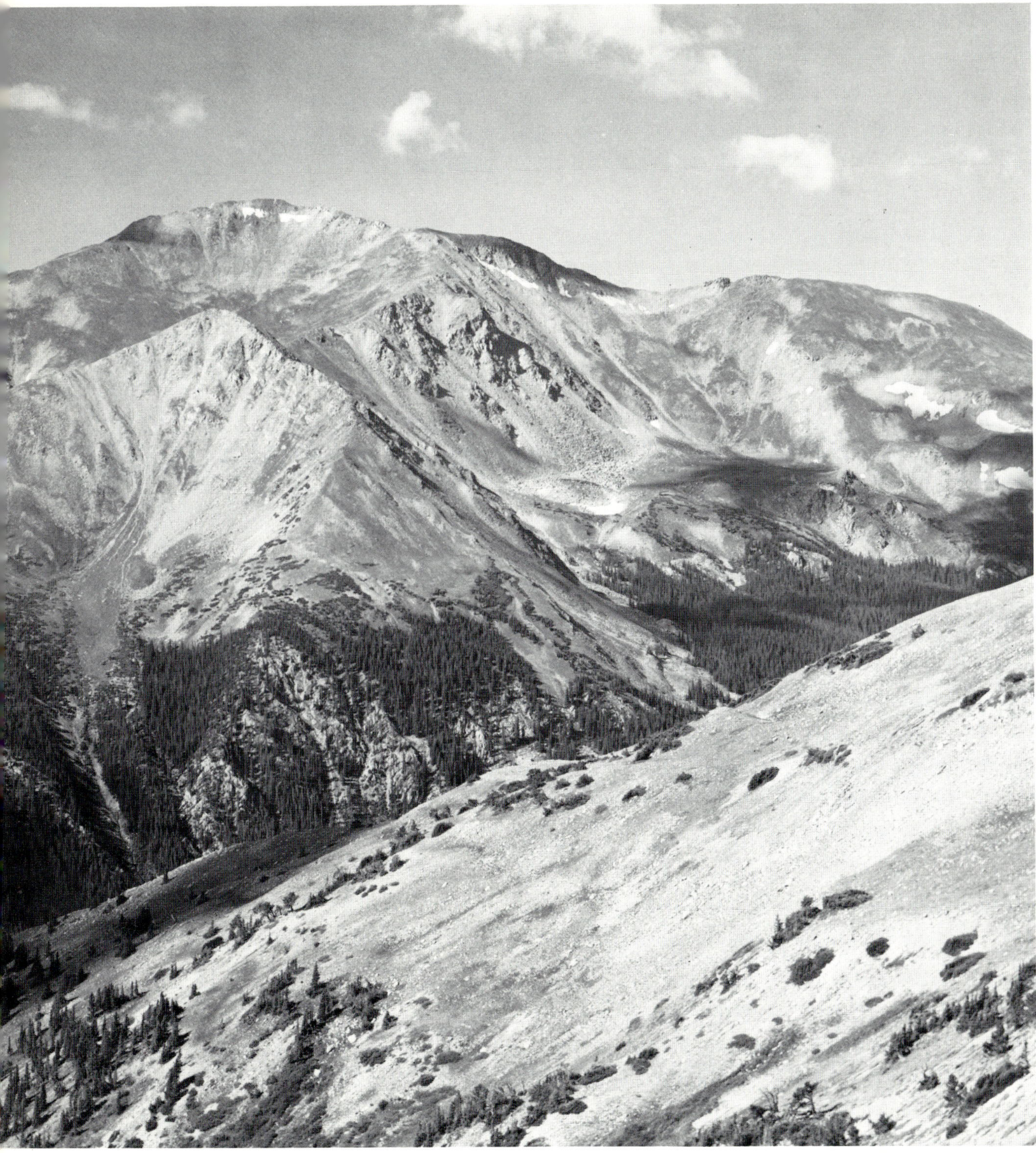

MOUNT PRINCETON
(Sawatch/Collegiate Range) 14,197 feet
18th highest

Although 17 Colorado peaks stand taller, Mount Princeton outdoes many by its beauty and rich history. In 1872, H. A. Merriam and E. W. Keyes discovered silver above timberline; their find developed into the Hortense, one of the area's first rich silver-producing mines. The mine operated profitably for many years before the ore gave out; that era of Mount Princeton's history has been immortalized by George Willison's classic *Here They Dug Gold.* Though the Hortense is gone, much mining is still being done on Princeton's slopes and along Chalk Creek, which separates this mountain from Mount Antero. Once an Indian gathering place, Mount Princeton Hot Springs became a resort during the great silver days and was known as the "millionaire's playground." Currently experiencing another tourist boom, the Hot Springs are located along Chalk Creek, just south of the peak.

Impressive white chalk cliffs on the southern slopes were the source of the peak's original name, Chalk Mountain. A 200-year-old treasure, which legend says was buried somewhere on those cliffs, is still being sought today. A band of conquistadores is said to have raided an Indian village of all its valuable trinkets while the braves were away hunting. When the Indians pursued them into the cliff region, the Spaniards were forced to lighten their load by hiding the two calfskins filled with booty. The treasure site remains a mystery even today, since the Spaniards were eventually caught and slaughtered by their pursuers. The cliffs make treacherous climbing, yet many people still believe in the story enough to risk the dangers. Several have died in the attempt. Serious climbers are also willing to tackle the cliffs for the challenge presented, rather than in quest of treasure.

The peak was measured by both F. V. Hayden and George M. Wheeler, Hayden's colleague and competitor in Western surveys. Wheeler, who had named it "Chalk Mountain," calculated its height as 14,041 feet. During the same period, Hayden came within a foot of its official height when he calculated it to be 14,196 feet.

Mount Princeton is the second highest peak in the Collegiate group, and was named by and for the Princeton University Scientific Expedition of 1877. William Libbey, Jr., a member of the party, climbed the mountain on July 17 of that year and estimated its height as 14,208.90 feet. Although there is no record of any Wheeler or Hayden team member having climbed the peak, it is possible that earlier prospectors made the climb before Libbey. Mount Princeton offers an interesting though not especially difficult climb, and provides a magnificent view in all directions.

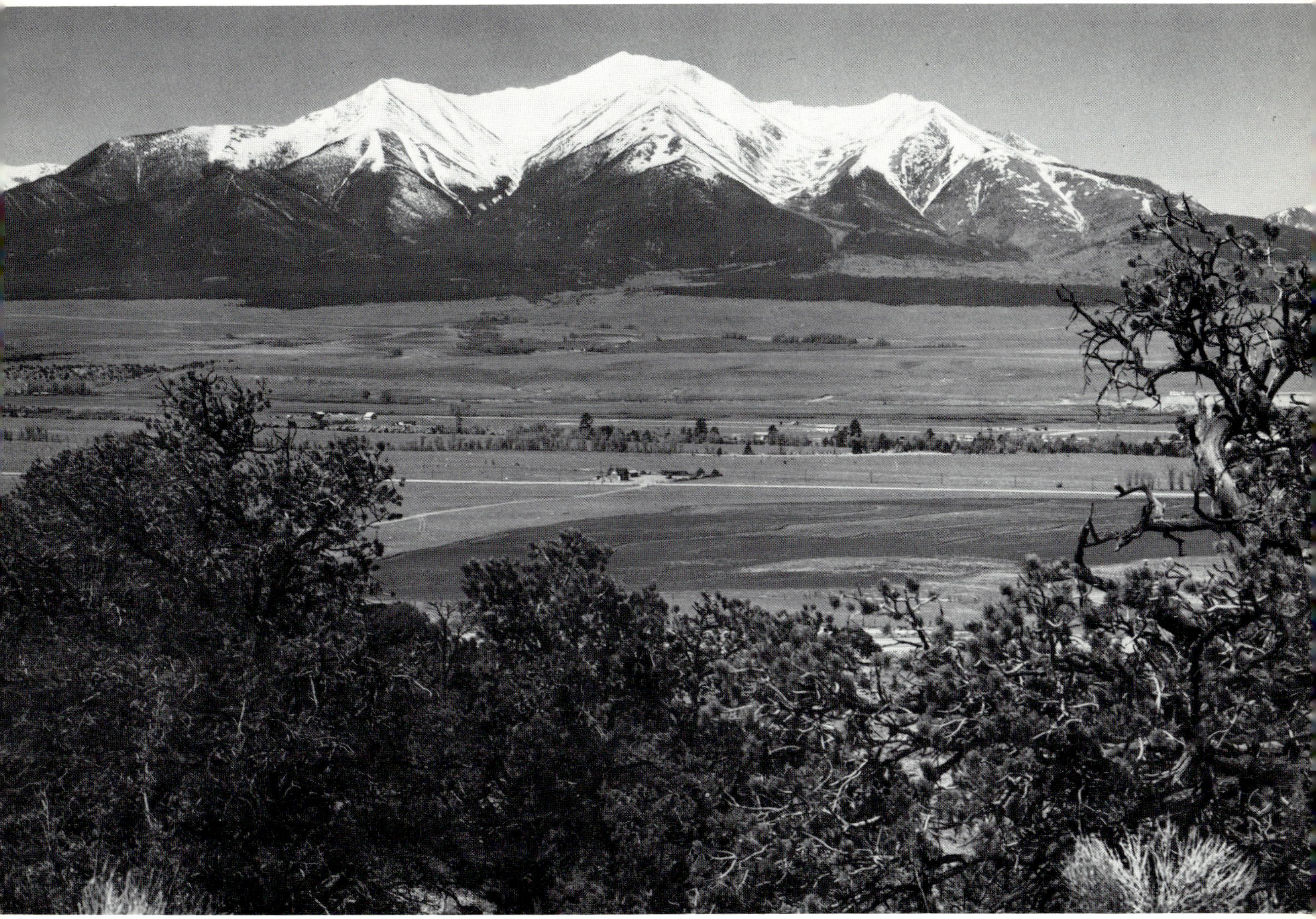

MOUNT ANTERO (Sawatch Range) 14,269 feet
10th highest

Mount Antero is the Colorado mountain most noted for gems and minerals and is the "highest mineral locality in North America." On Colorado Day, August 1, 1949, the Colorado Mineral Society climbed to near the summit of the peak, placed a bronze plaque on the granite cliff here, and proclaimed the creation of "Mount Antero Mineral Park." Antero is most noted for its aquamarine, rock crystal, and smoky quartz. Also found here is beryllium, phenacite, flourite, topaz, and mica. Gems found here, known to students, collectors, and dealers around the world, are generally of a blue to pale green color with some the prized deep blue of the Brazilian variety. Most of the best gems have been found near the top of the mountain.

Mount Antero has yielded many gems weighing more than 12 carats. Also, the mountain produced one of the largest clear rock crystals ever found. It was put on display at the 1893 Columbian Exposition and was valued at thousands of dollars. A crystal found by Edwin W. Over, Jr. in 1932 was seven inches long and more than one inch across, and is now in the Harvard Mineralogical Museum. Many other gems found here are on display in museums around the world. Most of the gems, however, are in jewel boxes.

Although it is believed that some gems were found high on the peak by prospectors, and earlier yet by Indians, the first recorded discovery was made by Nathanial Wanemaker around 1885. He built a small stone cabin in the amphitheatre on the south side of the mountain, about 800 feet from the summit. He lived alone here for years, mining what gems he could. It is reported that his first year's find was valued at about $600. Others came and by the 1890's $5,000 and more in gems were found.

Most of the early gems discovered here were sent to Germany for cutting and then returned to Colorado to be set and sold to tourists. Prospect holes covered the mountain, particularly near the summit. Perhaps most of the best gems have been taken from Antero, but it is still a favorite hunting ground for treasure hunters. Some good finds have been made in recent years.

The mountain was first noted in history when in 1806 Zebulon Pike ate Christmas dinner at its foot. The mountain was named for another Ute Chief who, with Shavano, worked to make peace between the whites and the Indians. Mount Antero is often called Antero Peak.

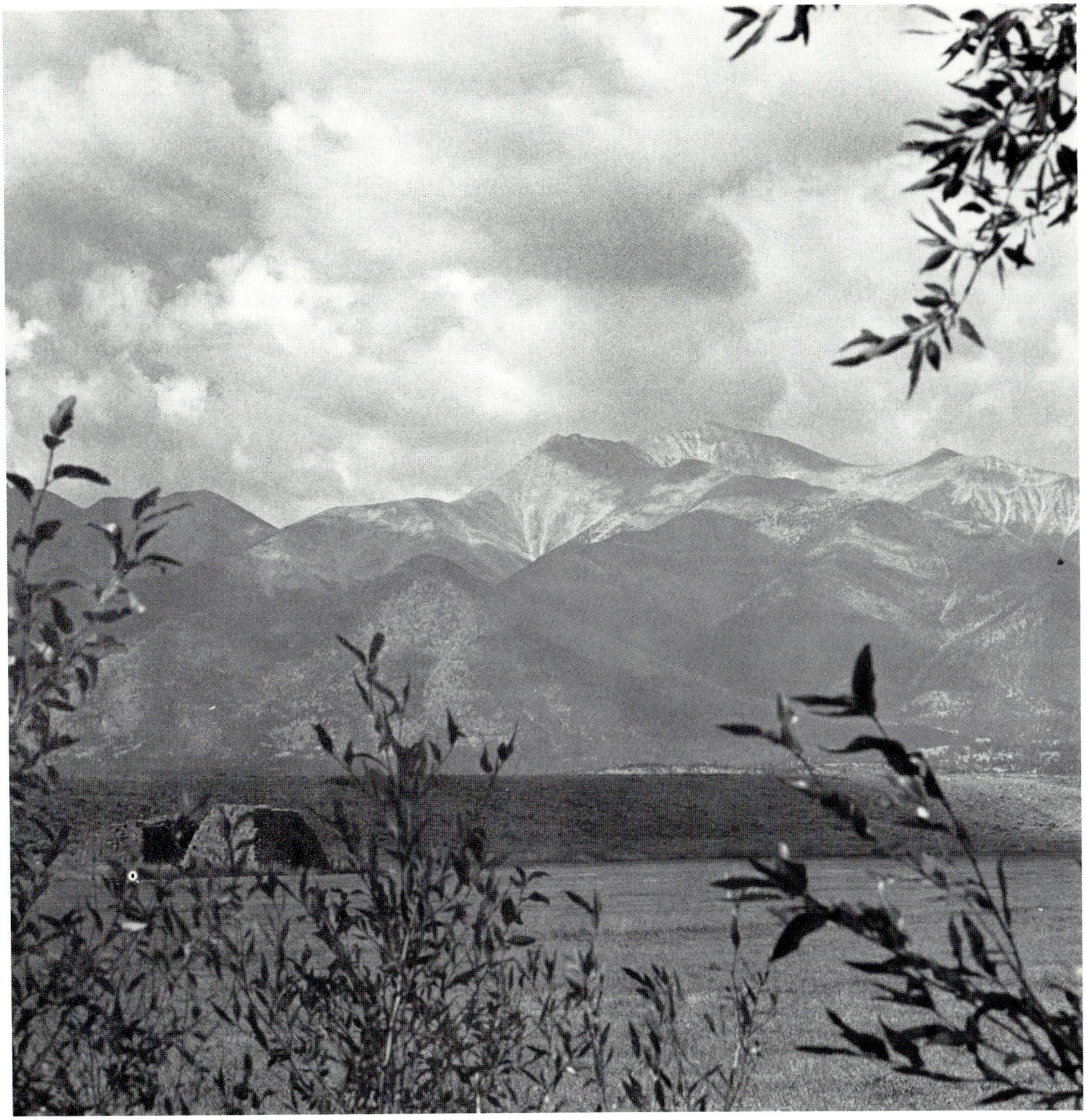

MOUNT SHAVANO (Sawatch Range) 14,229 feet
17th highest

In the late spring and during most of the summer, snow left in deep crevices high on this peak forms a figure with outstretched arms. The "Angel of Shavano," as the figure is called, can be seen from as far away as Salida and has become known around the world. There are several popular legends about how the "angel" came to be there.

One concerns the mountain's namesake, Chief Che-Wa-No, leader of the Uncompahgre Utes. Over the years, the chief had grown to love and respect the well-known white scout, Jim Beckwourth. In 1853, when Beckwourth was fatally injured in a riding accident, Chief Che-Wa-No, or Shavano, came to the mountain to pray for his dying friend's soul. Since then, every spring at the time the chief originally prayed at the foot of the peak, the snowy Angel of Shavano returns to signal that the Indian's prayers have been answered.

Another familiar legend concerns an Indian princess and her love for her people. Many moons ago, widespread drought was driving the Indians from the land. The princess came to kneel at the foot of the peak and prayed for rain to end the drought. The Indian God of Plenty beckoned to the princess to sacrifice herself so that her people might survive. Each year thereafter, the princess — the Angel of Shavano — reappears, and weeps once more for her people. Her tears — the melting snow — fall on the land below and make it fertile.

Another story concerns a naughty young goddess, or Indian princess, who had so angered the gods that they turned her to ice. Her repentance, or perhaps the drought below, moved her to tears, and she redeemed herself by breaking the drought.

Yet another legend involves the love of two young Indian chieftains, Stone Face and Little Drum, for Corn Tassel, an Indian princess. According to Indian law, the two had to fight to the death of one of them in order to determine who would win her hand. Corn Tassel's true love, Little Drum, was on the losing end of the struggle. Just before his own death, he shot an arrow through the heart of Corn Tassel, believing that if he could not win her, no one else would. Troubled days followed her death. Drought forced the tribe to leave the land where they had once prospered. Years later, the image of Corn Tassel showed itself on the mountain. By her tears the land again turned green, and the tribe was revived in both spirit and prosperity.

There are also many stories of how the sudden view of the Angel pulled many a troubled soul through hardship or was able to inspire them to great accomplishment.

Che-Wa-No ranked not far below Chief Ouray in working with the white man to keep the peace.

When a tribe of the Tabeguache Utes, led by Chief Kaneache, went on the rampage in 1867, Kit Carson sent Shavano out to bring Kaneache back and end the uprising. Shavano did so in short order, and the following year he was sent to Washington, where great honor was shown him for his heroism. Named chieftain by Ouray, Shavano worked closely with the whites in putting out brush fires and establishing and maintaining peace with the Indians. He was one of three Indians accompanying General Adams to recover the whites captured during the Meeker Massacre in 1879. Despite his work, however, he and the others were deported to Utah in 1881.

It is fitting that a Fourteener bear the name of Shavano. Curiously, however, a nearby mountain which is named for Chief Ouray, the greatest Ute of them all, is not a Fourteener.

Not a difficult climb, Shavano was no doubt topped by prospectors. There was mining for both gold and silver done all over the mountain, and much of it close to the summit. Remains of the old mining town of Shavano are about halfway up the mountain. The first publicized climb up the peak was made in 1888 by mountaineer Charles Fay. Probably the only remarkable thing about Fay's climb was that it was begun at 2:30 A.M.

TABEGUACHE MOUNTAIN
(Sawatch Range) 14,155 feet
26th highest

This is one of the Fourteeners that has been called both "peak" and "mountain" — take your pick.

The history of neglect of this mountain parallels that of Mount Oxford. It was noted but unnamed for many years, and apparently was not welcomed into the Fourteeners until about 1930. Then, even after it was recognized, it was inaccurately listed as being an even 14,000 feet until very recently. It gained almost as much stature as Mount Oxford in subsequent surveys.

The mountain was finally named by the Colorado Mountain Club for a branch of the Ute family that once roamed the area. In 1867, some of the Tabeguache went on the rampage, led by Chief Kaneache. Kit Carson asked another Ute Chief, Shavano, to capture Kaneache, and he succeeded in doing so. Both Shavano and Antero were Tabeguache (pronounced tab-i-wash) chiefs and have had Fourteeners named after them.

The summits of Tabeguache and Shavano are about 1¼ miles apart, and there is a depression of about 400 feet between them. The two are generally climbed together with the most popular starting place being the Shavano Camp Grounds.

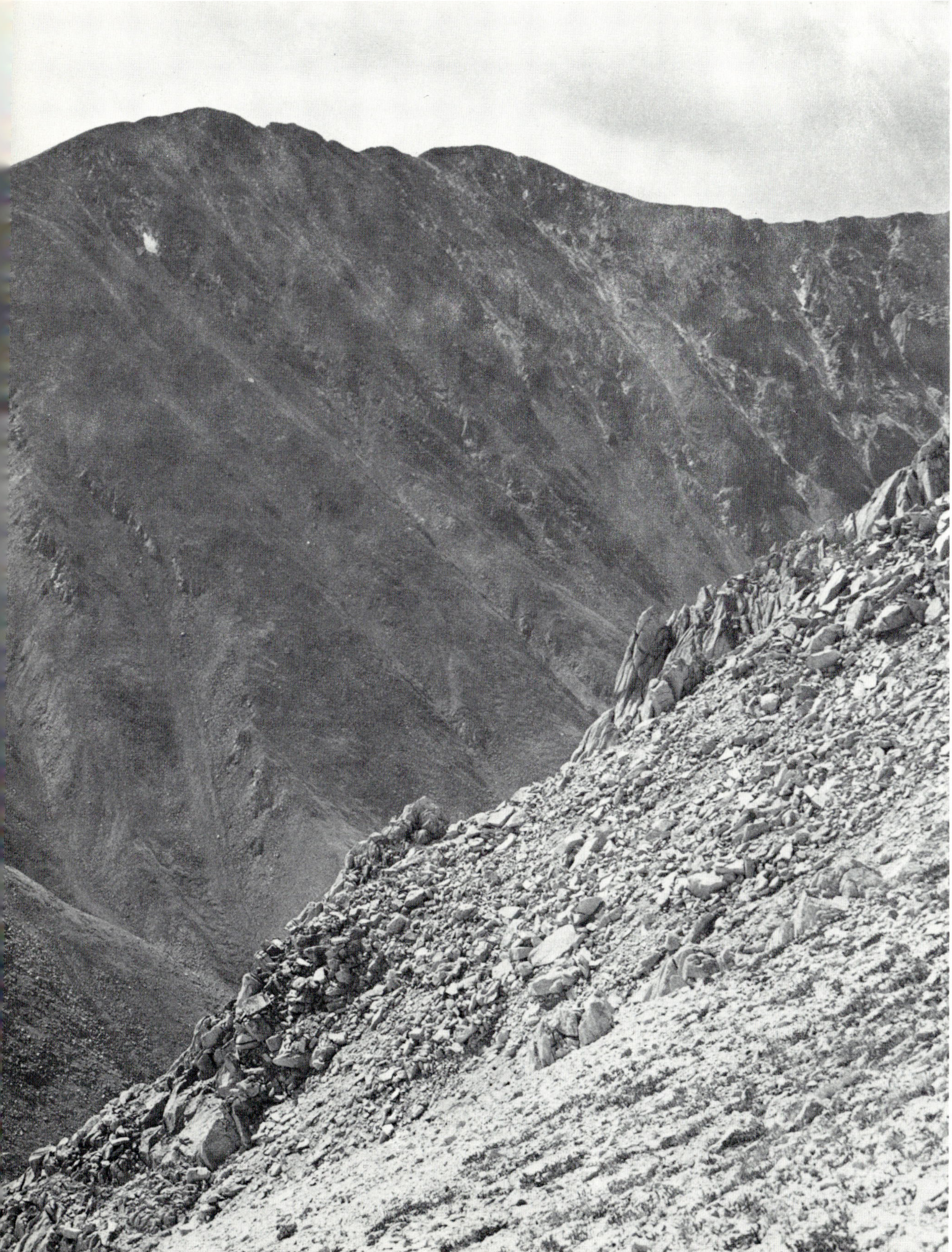

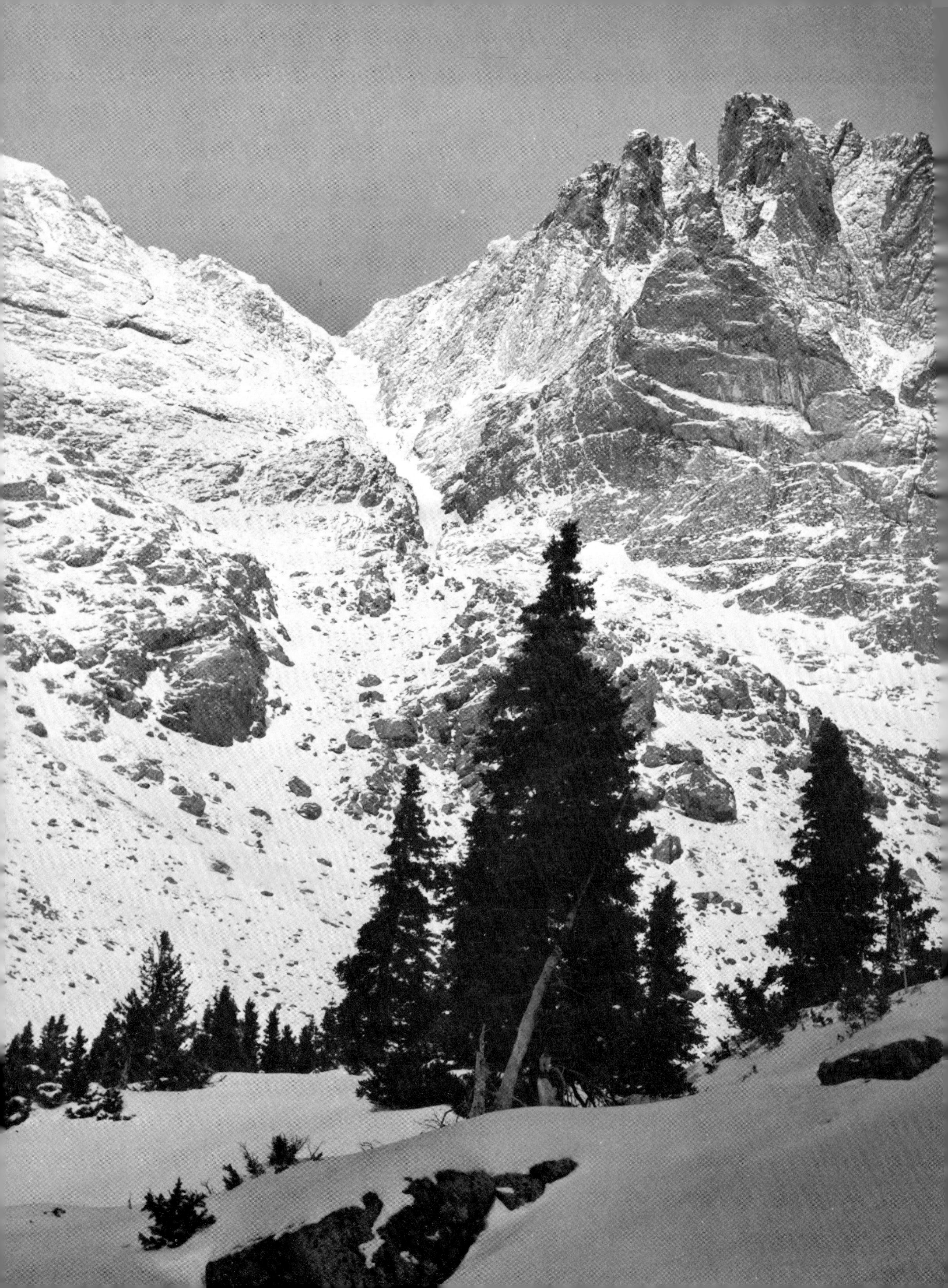

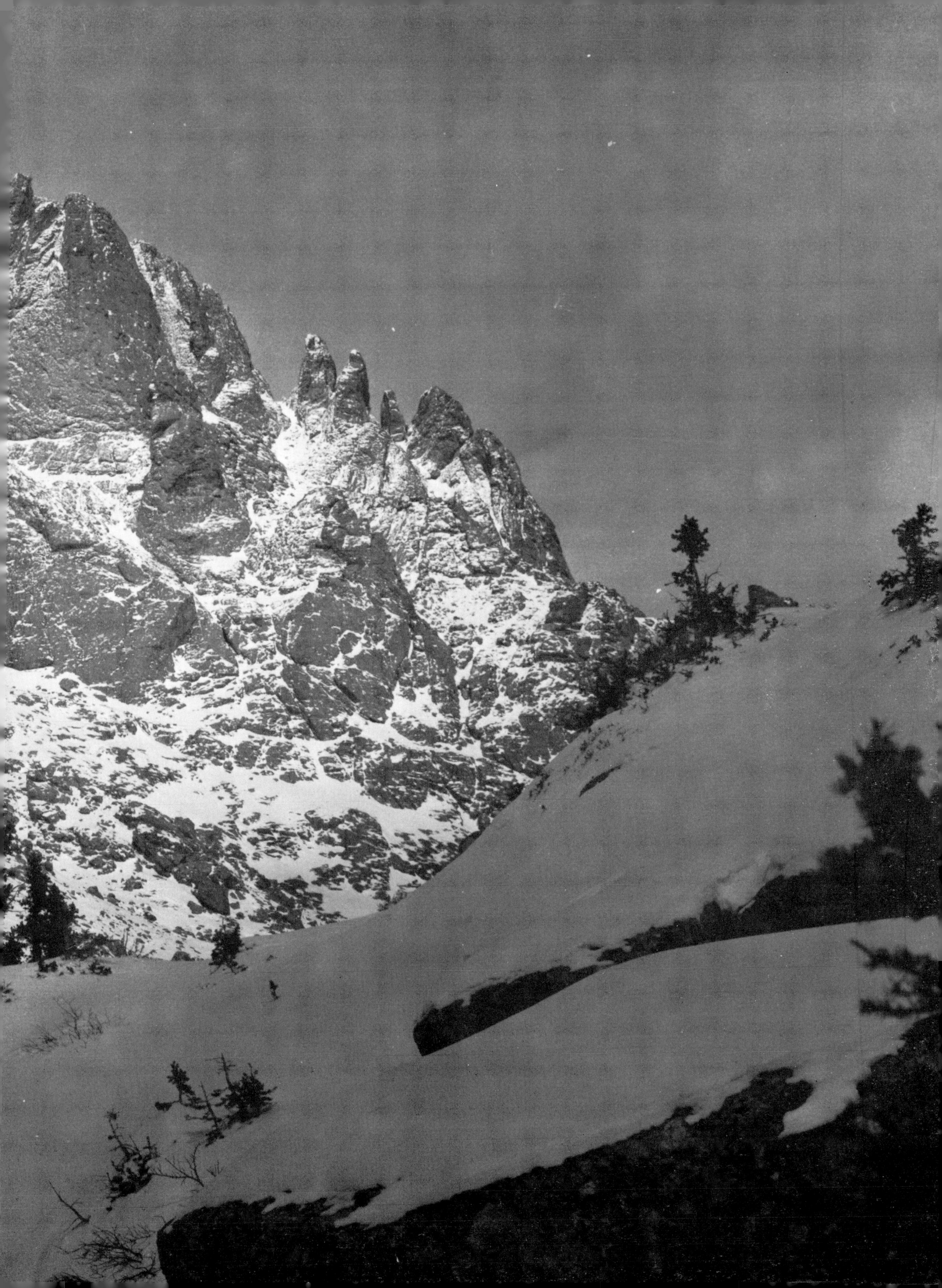

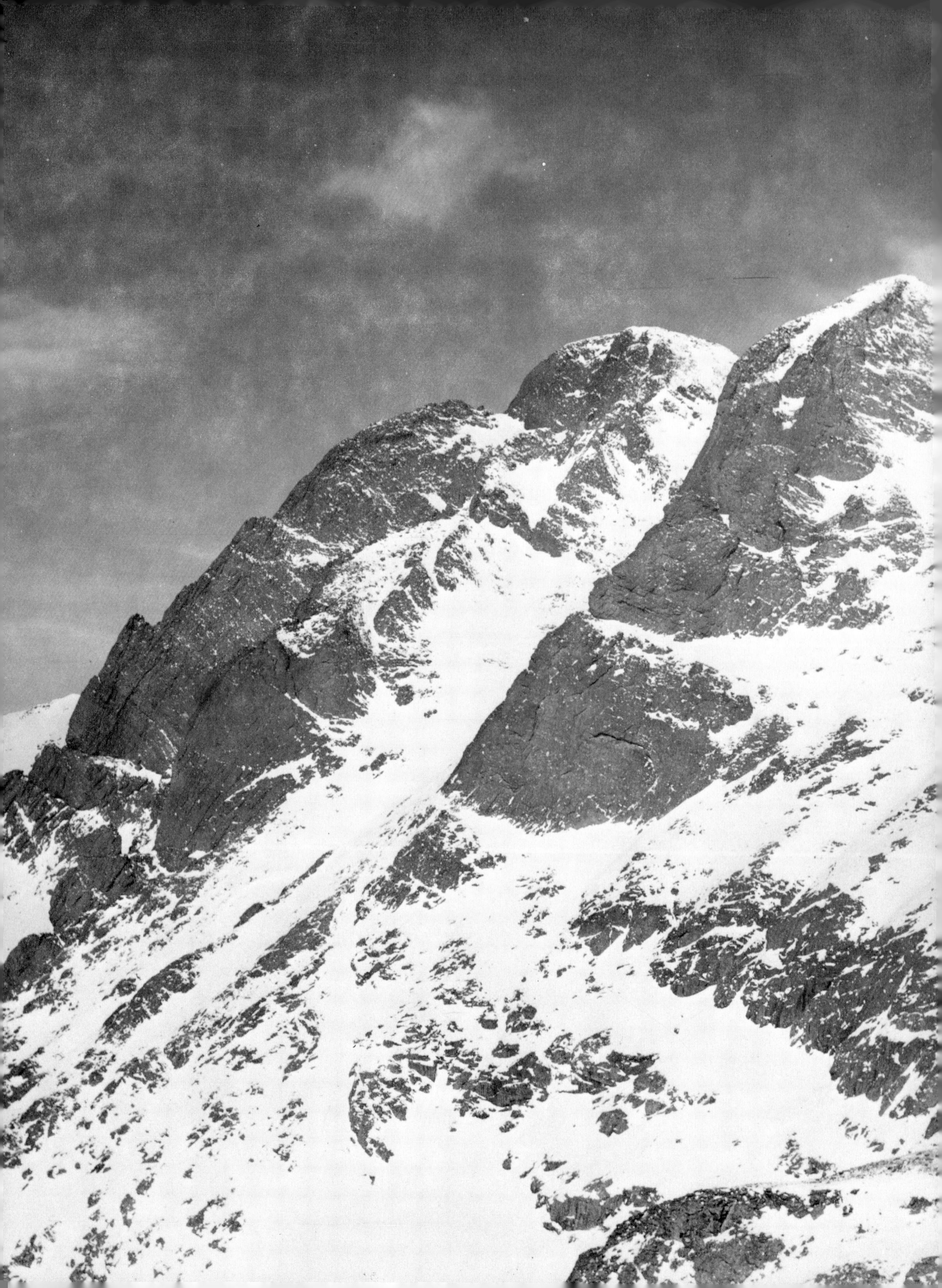

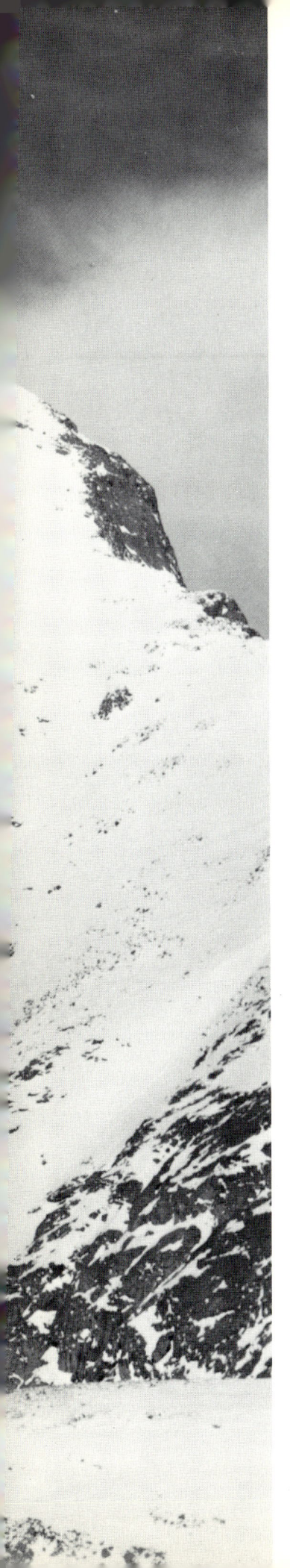

KIT CARSON PEAK
(Sangre de Cristo Range) 14,165 feet
23rd highest

This is another peak that has "grown much" in recent years, gaining 65 feet and rising from 32nd to 23rd highest among the Fourteeners.

The mountain was mapped by the Hayden Survey in 1874 and believed named for the famous scout at that time. Earlier, it and the nearby Crestone peaks were called "Trois Tetons," or Three Tetons, for their similarity to the mountains in Wyoming. Kit Carson Peak was also called "Haystack Baldy" for its appearance, and "Frustrum Peak" for a member of the Wheeler Survey team.

Some claim that Kit Carson once had a cabin on the south slopes of the mountain. It is possible that the cabin in which he stayed was built by another mountain man, "Moccasin Bill" Perkins. Perkins was an early hunter in the region who always wore moccasins to more quietly stalk his game. Moccasin Bill was also quite a story teller, and was the source of many lost mine stories and other legends.

Kit Carson Peak and the Crestones were among the last Colorado Fourteeners to be climbed. Many unsuccessful attempts were made on the mountains, some ending in tragedy. For many years these peaks were believed unclimbable, until negotiated finally by members of the Colorado Mountain Club in 1916. They have been climbed several times since, but only by skilled alpinists.

HUMBOLDT PEAK
(Sangre de Cristo Range) 14,064 feet
37th highest

This peak, nestled among the Crestones, was named for Alexander von Humboldt, noted geographer, explorer, and world traveler of his day who climbed mountains throughout the world. It is believed the peak was named by German settlers in the Wet Mountain Valley, many of whom worked at the Humboldt Mine located on the slopes of this mountain.

Apparently, the peak was first climbed by survey crews in 1883, many years before its more challenging neighbors were scaled. Unlike the Crestones and Kit Carson, Humboldt is an easy "walk-up". The climber is rewarded by exciting views of the neighboring mountains and the valley below. South Colony Lake is a popular base camp for the climb.

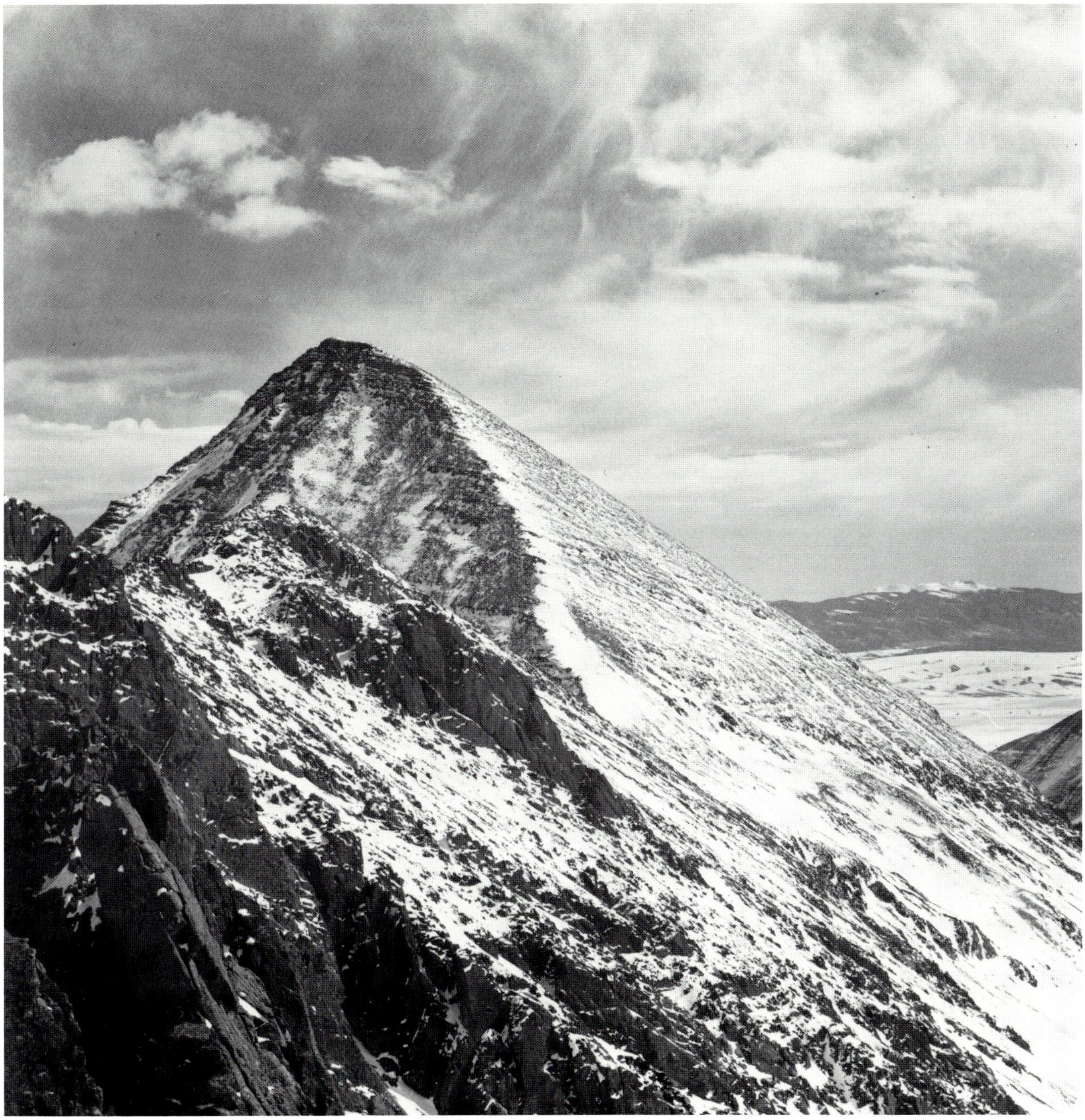

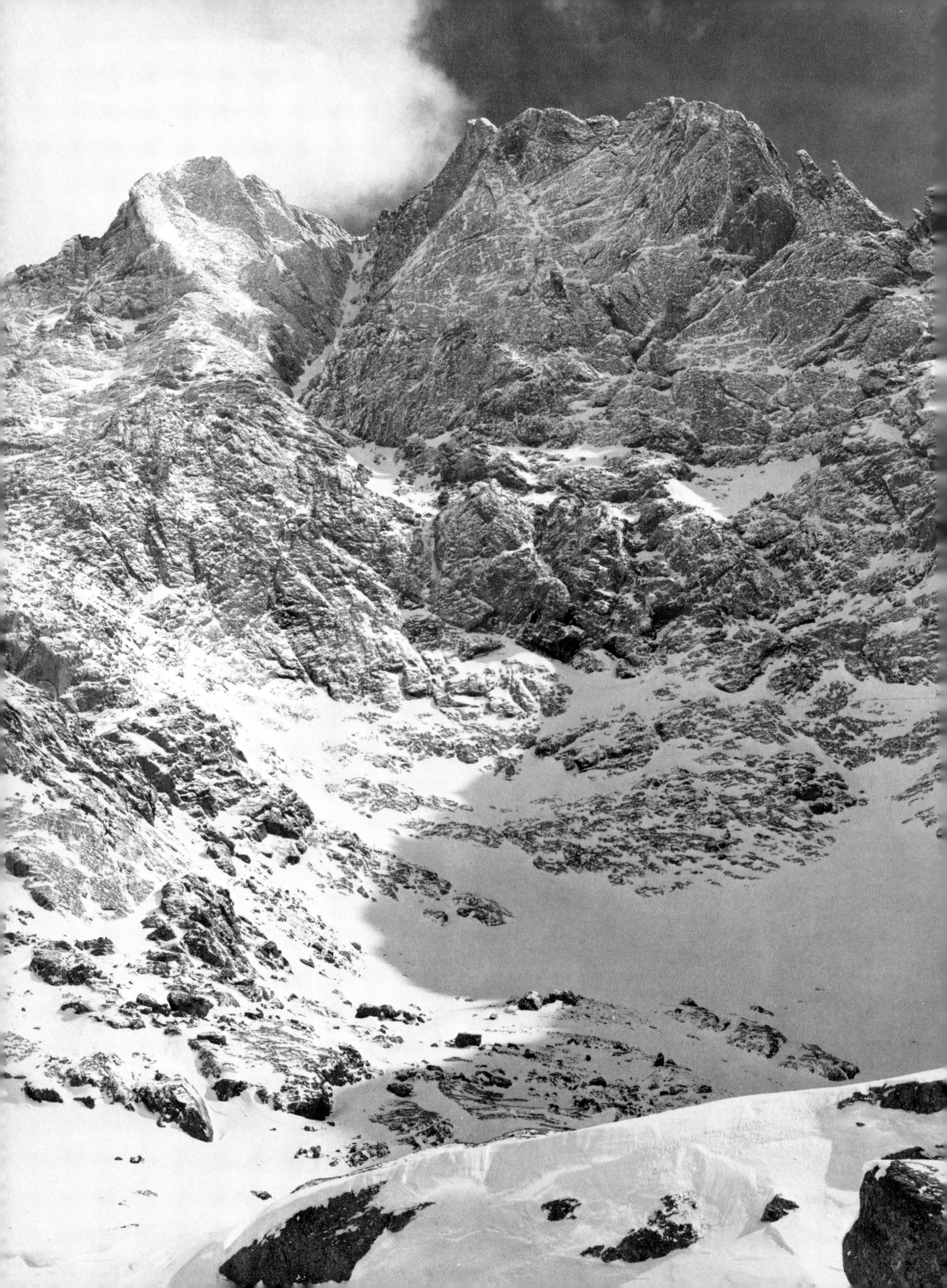

CRESTONE PEAK
(Sangre de Cristo Range) 14,294 feet
7th highest

This and its two neighbors, Crestone Needle and Kit Carson Peak, form one of the most impressive mountain sights in Colorado, and they pose a severe challenge to dedicated mountain climbers.

Crestone Peak is believed to have been the last climbed of all Colorado's Fourteeners. It was first officially negotiated in 1916 by Albert R. Ellingwood and Eleanor S. Davis. Enthusiasts who set out to climb every one of the Fourteeners generally save Crestone for last, since a climber needs every ounce of experience he can muster in order to conquer this peak. Today, experts usually climb this and Crestone Needle together, going from one peak to the other via a steep, spectacular ridge.

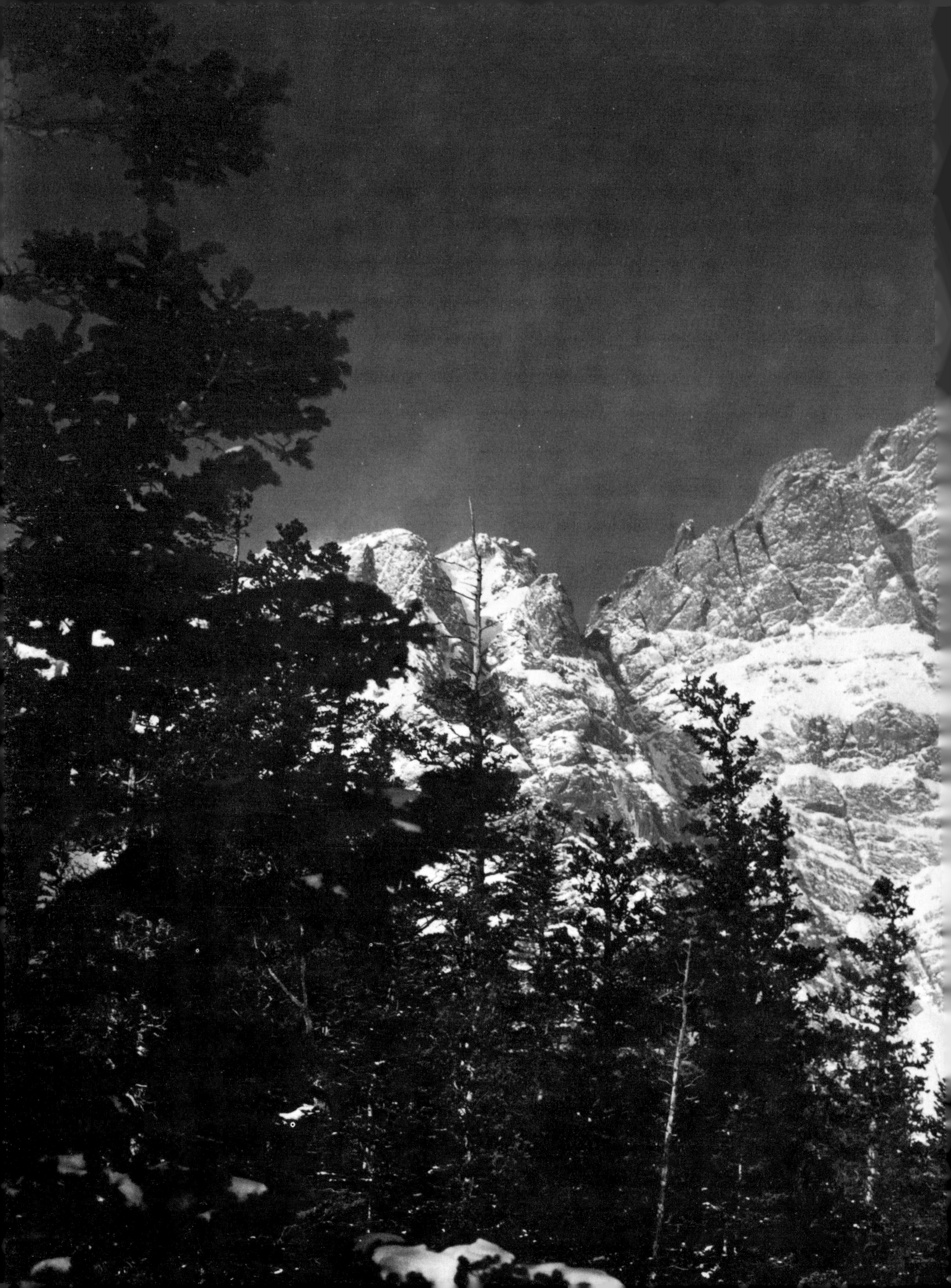

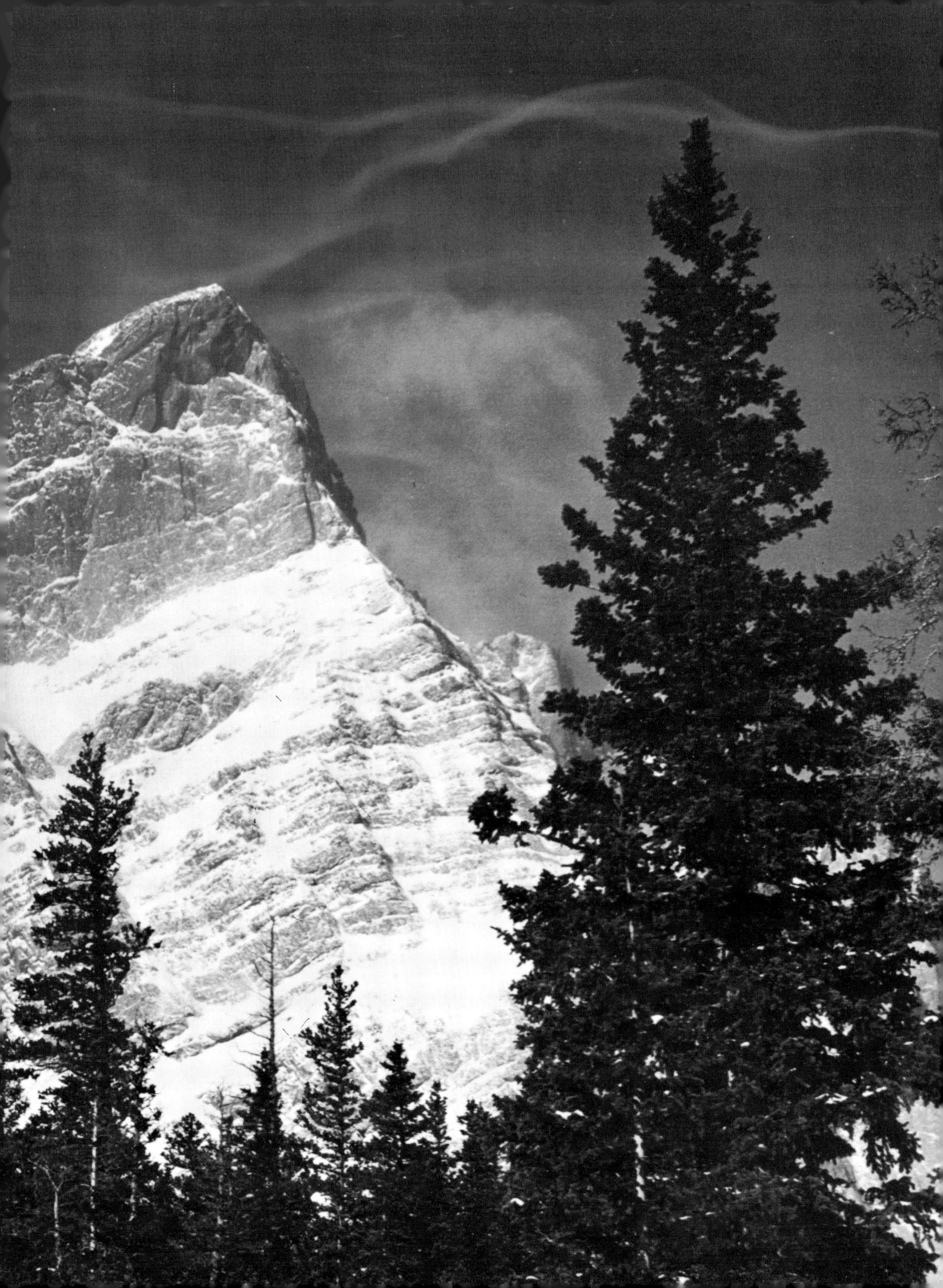

CRESTONE NEEDLE
(Sangre de Cristo Range) 14,191 feet
21st highest

Crestone ("crestón") is a Spanish word with several meanings from which to choose: cockscomb, large crest, or outcropping of ore. The jagged silhouette presented by both this and Crestone Peak suggest that they were named for a rough resemblance to a cockscomb. Early names by which this peak and its neighbors were known included "Trois Tetons" or Three Tetons, "The Needles," and "Spanish Crags."

Crestone Needle was first climbed only a short time before the last hold-out, Crestone Peak. Each year a few experts are able to climb the two of them together.

MOUNT LINDSEY
(Sangre de Cristo Range) 14,042 feet
42nd highest

Until 1954 this mountain was known as "Old Baldy." Since there are no less than 10 mountains in Colorado named "Baldy" and 15 named "Bald Mountain", it was altogether proper that "Old Baldy" was renamed for one of the state's most avid mountaineers.

Malcolm Lindsey grew up in nearby Trinidad and climbed Baldy frequently as a youngster. He was legal counsel for the city of Denver from 1925 to 1937 and was Denver's city attorney for the next decade. He held many other official and advisory posts and was one of the state's foremost experts on water problems and municipal law.

Lindsey's heart was always in the highlands, however. He joined the Colorado Mountain Club in 1922 and held many posts in the organization during the next few decades. He led many a youth group on mountain climbing expeditions, communicating to hundreds of youngsters his love for the mountains. His favorite mountain was always Old Baldy, and it is doubtful that any other man has climbed the mountain as frequently or knew it as well.

The name change was recommended by the Colorado Mountain Club and approved in 1953, shortly after Lindsey's death. The mountain was officially dedicated with impressive ceremonies on July 4th, 1954. Participating in the ceremonies were the Colorado Mountain Club, the Denver Bar Association, and the Colorado Episcopal Church, another group to which Lindsey dedicated his energies.

A marker was placed at the foot of the peak in a roadside park off Highway 160, three miles east of Fort Garland, in May of 1955. The marker was stolen within the next month.

Many oldtimers still reject the name change, and they still call it Old Baldy. It was probably first named Baldy by early miners, who noted that much of the mountain was bald of trees and who also, apparently, liked to name mountains "Baldy" or "Bald Mountain."

It is not a difficult climb, except for some steep upper ridges. This and the fact that it is off the beaten track make the peak one of the least popular climbs among the Fourteeners, despite the efforts of Lindsey. Indians and prospectors certainly reached the summit earlier, but the first recorded climb was made by both the Wheeler and Hayden Survey teams in 1875. It was seldom climbed again until Lindsey came along. The peak is located, as the crow flies, about two miles east of Blanca, but it is often lost in the ridges and trails over the massif. No doubt many a mountaineer believed he had reached the summit of Mount Lindsey, after having actually climbed some other peak instead. The climb is usually made from either the north or the south.

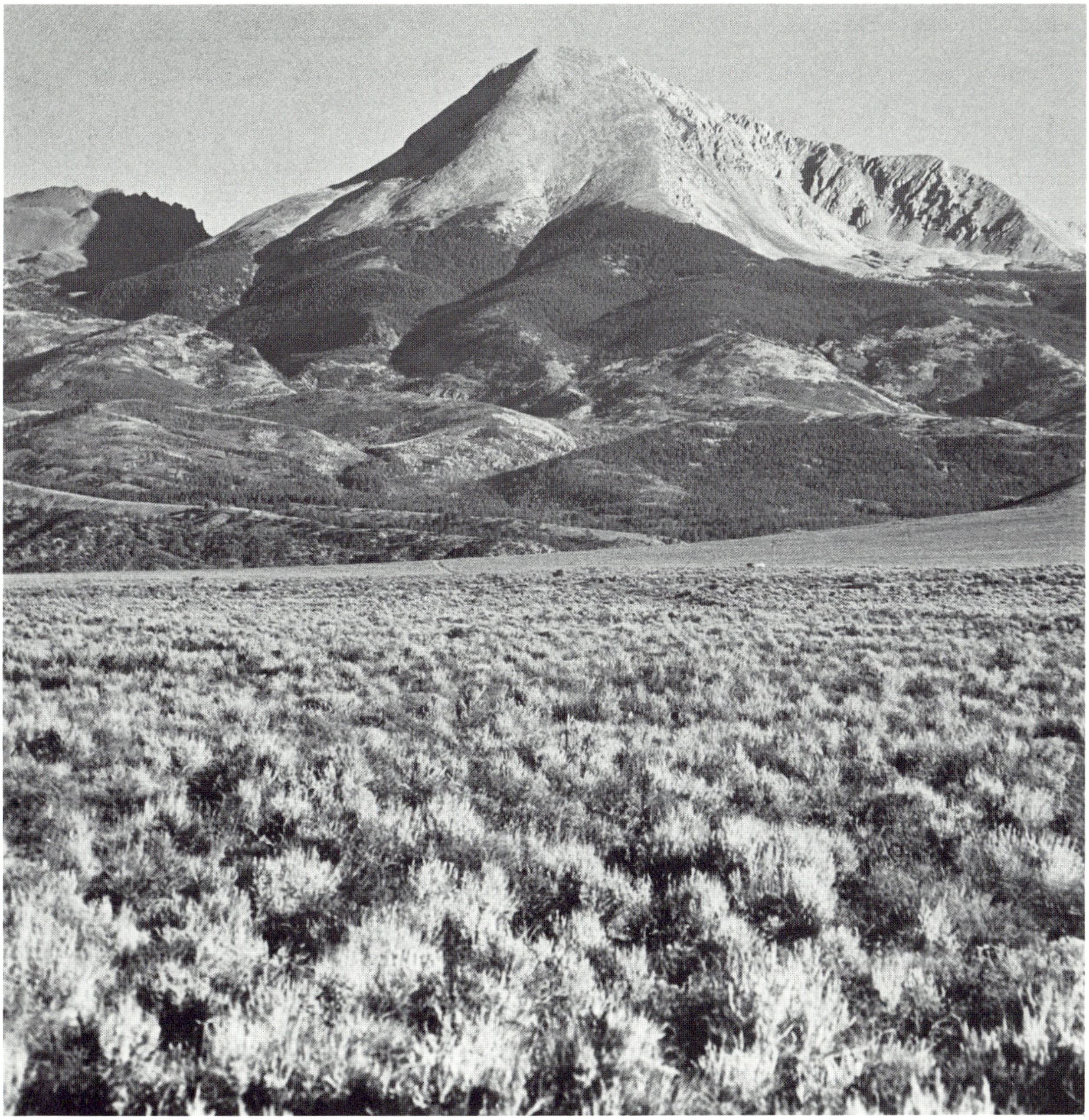

LITTLE BEAR PEAK
(Sangre de Cristo Range) 14,037 feet
43rd highest

Little Bear was at first called "West Peak," probably because it is west of Mount Lindsey, another satellite peak of Mount Blanca. Little Bear Creek, which runs on its slopes, gave this mountain its present name.

It was first officially climbed by Professors Fay and Edmunds in 1888. The peak is at the west end of a jagged 1½-mile ridge running from Blanca Peak. It is one of the more difficult climbing mountains in Colorado, recommended for experts only; however, one of the most manageable routes begins from the south. Little Bear also has satellite peaks, North and South Little Bears, with another difficult ledge connecting "Mama" Little Bear and North Little Bear.

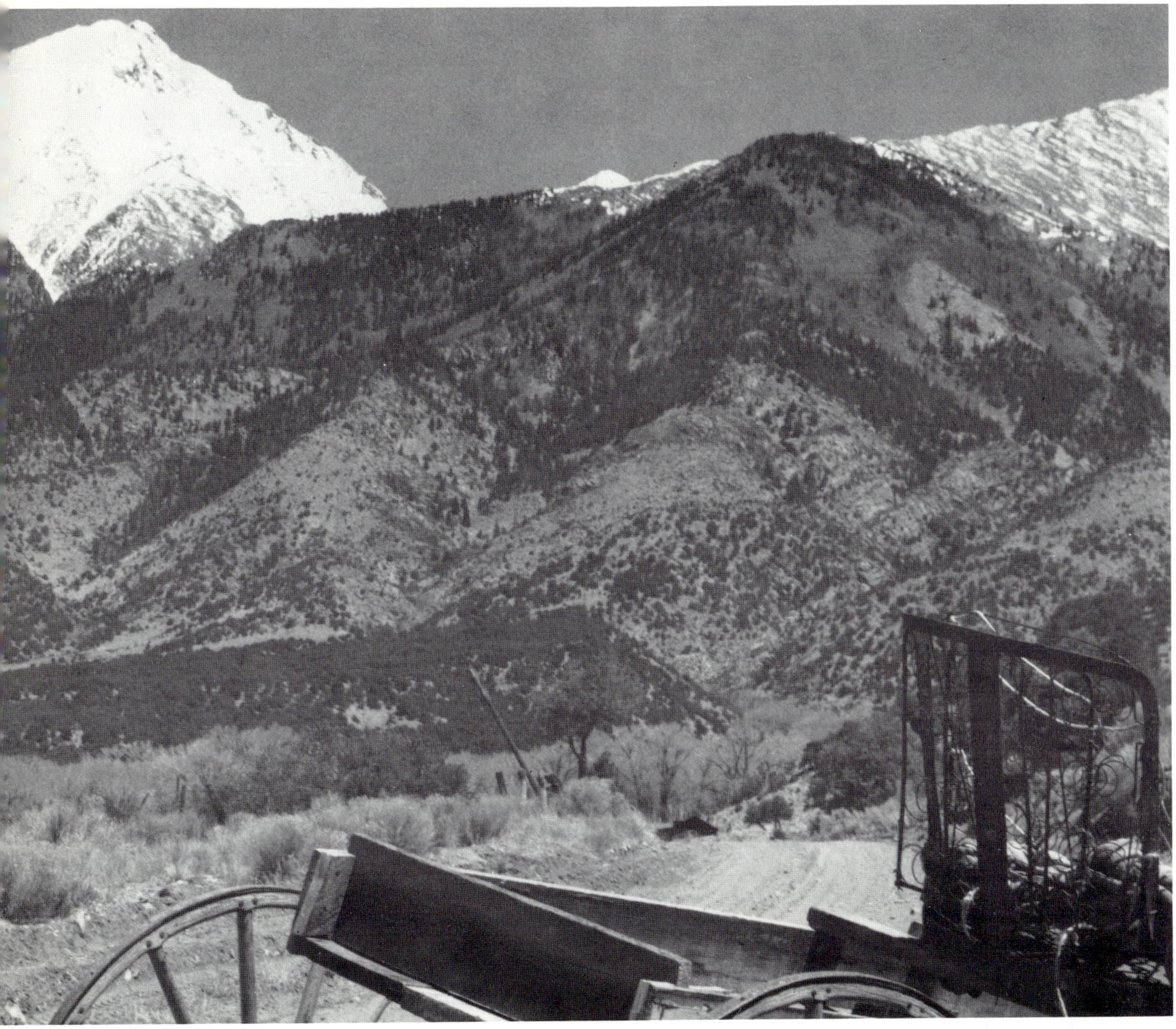

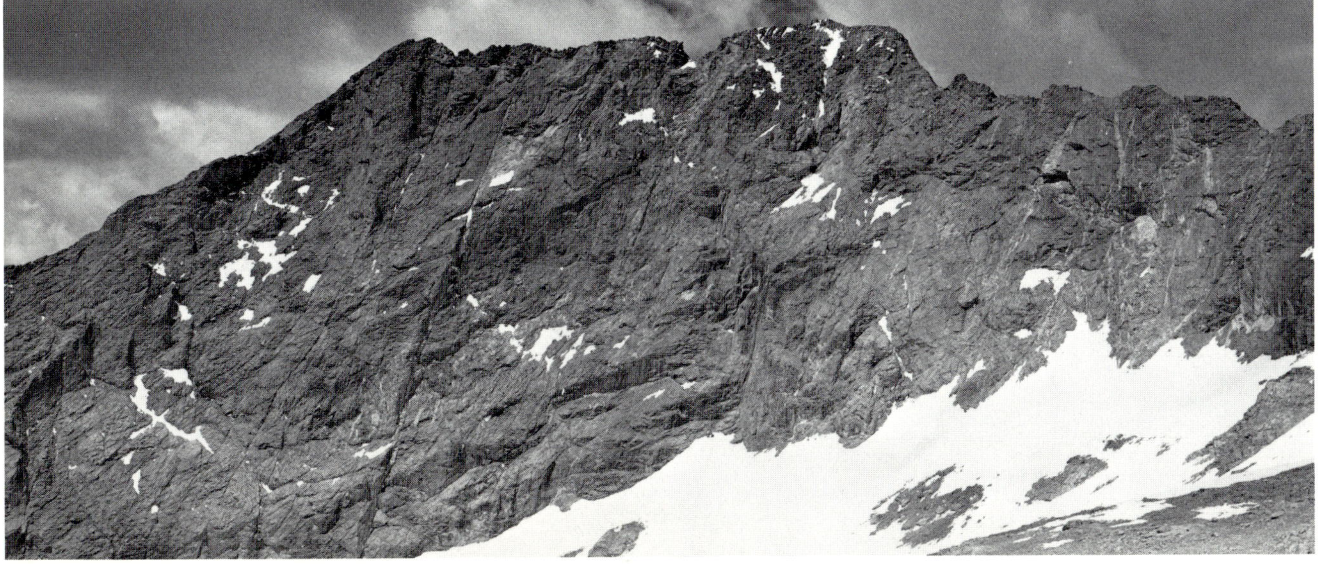

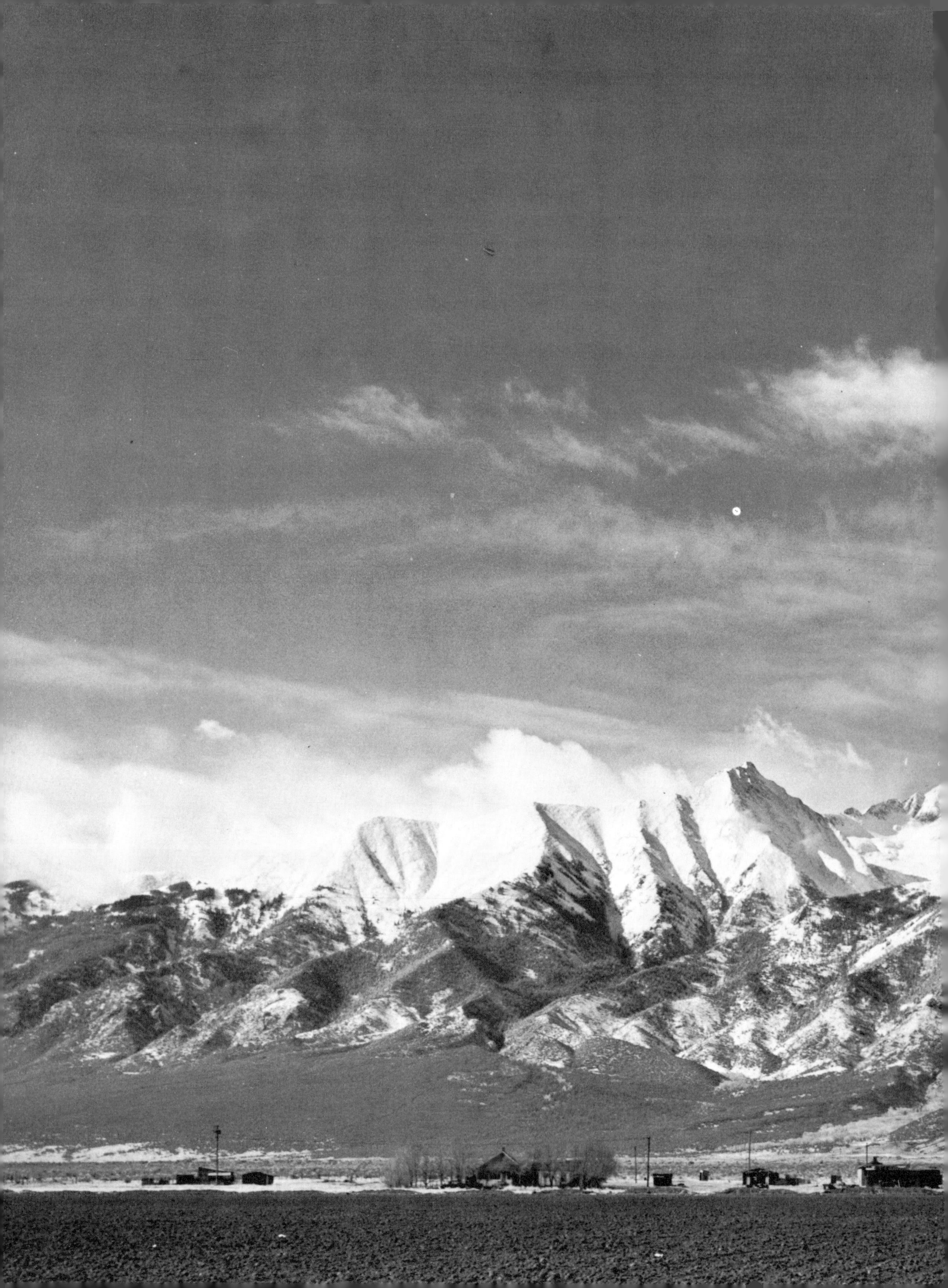

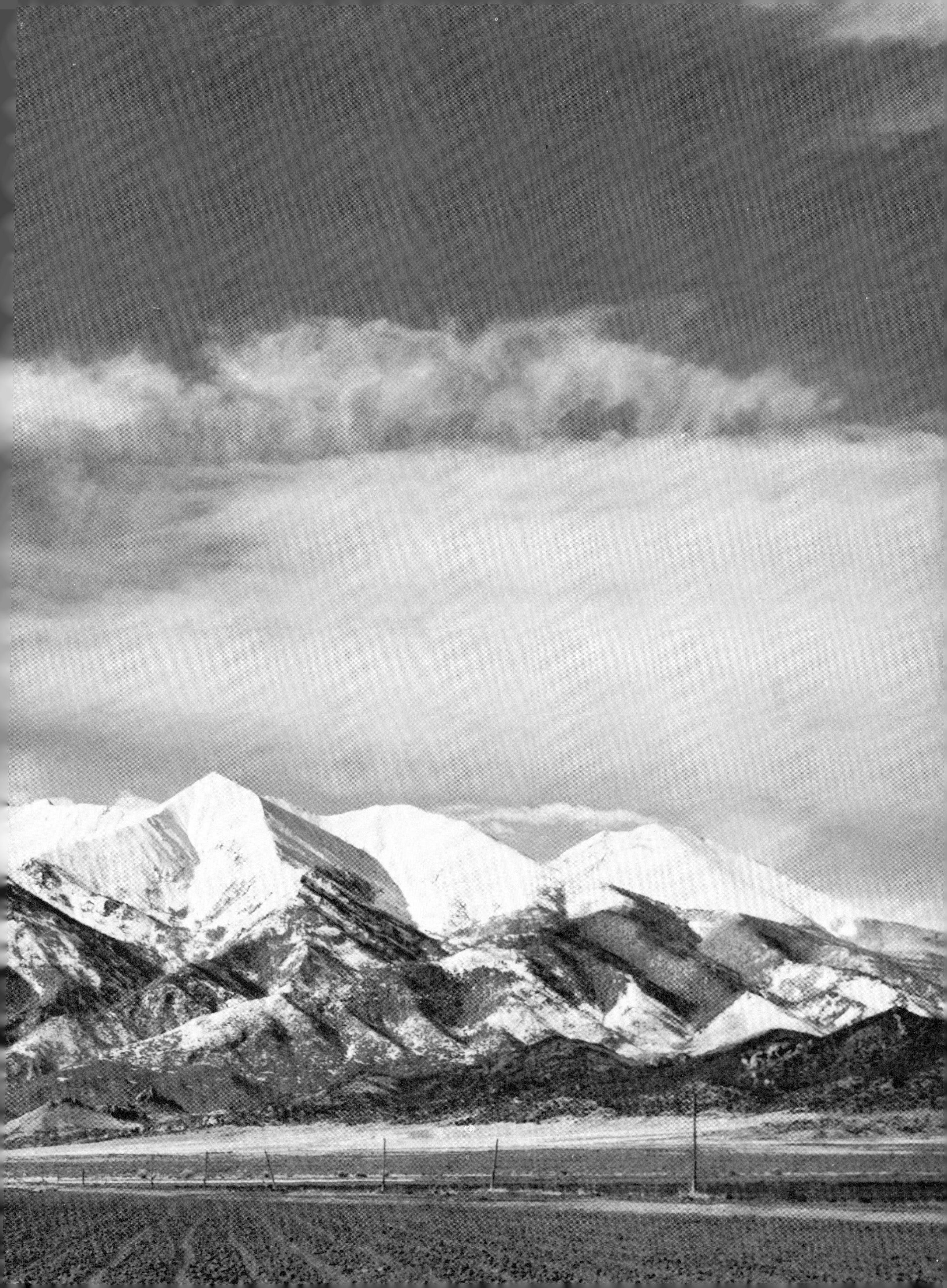

BLANCA PEAK
(Sangre de Cristo Range) 14,338 feet
4th highest

That this isn't the highest peak in Colorado — and even the United States (excluding Alaska) — is not the fault of its many promoters. Foremost among these is former Colorado playboy Donald E. Bennett. Bennett, with plenty of money and much leisure time, set out to make Blanca the highest and most heralded peak. Many early surveys and estimates showed Blanca to be the highest peak in the state and nearly the highest in the nation. Disappointed that later surveys had trimmed down the height of "his peak," Bennett set out to make amends. He used his influence and money in an attempt to force new surveys to prove his point. Realizing that that was a losing battle, Bennett next hired a work crew to pile rock atop the mountain in order to make up the almost 100-foot difference between Blanca and Mount Elbert, officially Colorado's highest. Still unable to impress anyone with this maneuver, he determined to erect a 50-foot aluminum flagpole atop the stones. Due primarily to high winds, Bennett's first flagpole attempt failed. The pole was finally erected, however, and its tip was long believed to be the highest point in the state. The fact is still being debated.

Bennett also publicized his peak in other ways. He made radio broadcasts, as well as telephone calls to important people, from Blanca's summit, transmitting through barbed wire running to a telephone exchange.

The controversy isn't over yet. Though Bennett's ingenuity brings him close to realizing his "impossible dream," it is doubtful that his tactics would ever receive official sanction.

One recent Geological Survey list of Fourteeners in Colorado allowed Blanca Peak 14,317 feet. But astericks fuel the fire. The footnote read: "Different sources of information give the elevation of Blanca Peak varying from 14,310 to 14,390. The U.S. Geological Survey has accepted 14,317 as being most reliable." Then a later survey listed it as 14,345 feet. This is the way it has been from the very first. Few other mountains in Colorado have been the subject of so much doubt. Hall's *History of Colorado* said the peak was 14,464 feet high, Colorado's tallest. This was the figure used by the *International Encyclopedia* and by early Colorado Mountain Club members. However, F. V. Hayden estimated that it was only 14,163 feet high. Then, the first survey by the United States Geological Survey in 1909 added 200 feet to that. There have been many other estimates by many other groups and organizations over the years; one of these would actually have excluded Blanca Peak from among the Fourteeners.

Some geologists believe that Blanca was certainly the highest peak in the state many millenniums ago, when, as a giant volcano, it blew its top to form the Sierra Blanca massif. Blanca is the highest point in what was left. Little Bear, Mount Lindsey, and other peaks outline the perimeter of the once gigantic mountain. Ridges run from Blanca in six directions, and the most prominent run to Little Bear and Mount Lindsey. One of the most majestic views found in the state is seen from the summit of Blanca.

The impressiveness of the mountain has made it prominent in Indian legend and a timeless landmark for travelers. Furthermore, the north face of Blanca cradles the southernmost glacier in North America. Zebulon Pike very probably saw the peak during his wanderings. Captain John W. Gunnison and many other early explorers also mention the mountain.

Blanca was first climbed officially by the Wheeler Survey in 1874, and then by the Hayden Survey the following year. Both parties discovered a curious fortification at the summit which was considered to be either an Indian eagle trap or lookout point.

Mountain climbers compare the climb with that of Matterhorn and other European peaks. The west ridge particularly is considered very dangerous, and only a handful of expert climbers have negotiated this route. Colorado Mountain

Club members have been able to scale the peak in winter.

In the 1890's many small mining camps bloomed on the mountain. The most well-known was the gold mining town of Commodore Camp. The camp was isolated in a canyon at the foot of the peak, and was, before the gold find, for many years merely a favorite spot for gathering wild raspberries.

Some early historians claim the mountain was named Blanca (Spanish for "white") for the prominent outcropping near the summit, but later mountaineers insist that the mountain was named for the snow habitually seen at its peak.

CULEBRA PEAK (Culebra Range) 14,047 feet
41st highest

Culebra is Colorado's southernmost Fourteener. It is a loner, located far from the other Fourteeners.

Culebra is the only Fourteener to have been owned by one man. The mountain once belonged to Delfino Salazar, who, until his death in 1958, operated the century-old general store in San Luis. The mountain was a part of old Spanish land grants under Queen Isabella of Spain. Salazar eventually bought out all the heirs and had himself a mountain, one of the largest and most unique ranches in Colorado.

Culebra means "snake" in Spanish, and some say the mountain was named for the many snakes in the area. Others insist there are not many snakes on the peak, and that the mountain was named for the serpentine curves of the river on its slopes.

The peak has long been a landmark. It was shown on Pike's map, published in 1810, and again on Alexander von Humboldt's map the following year. Culebra was first climbed officially in the mid-1870's by survey teams charting the region.

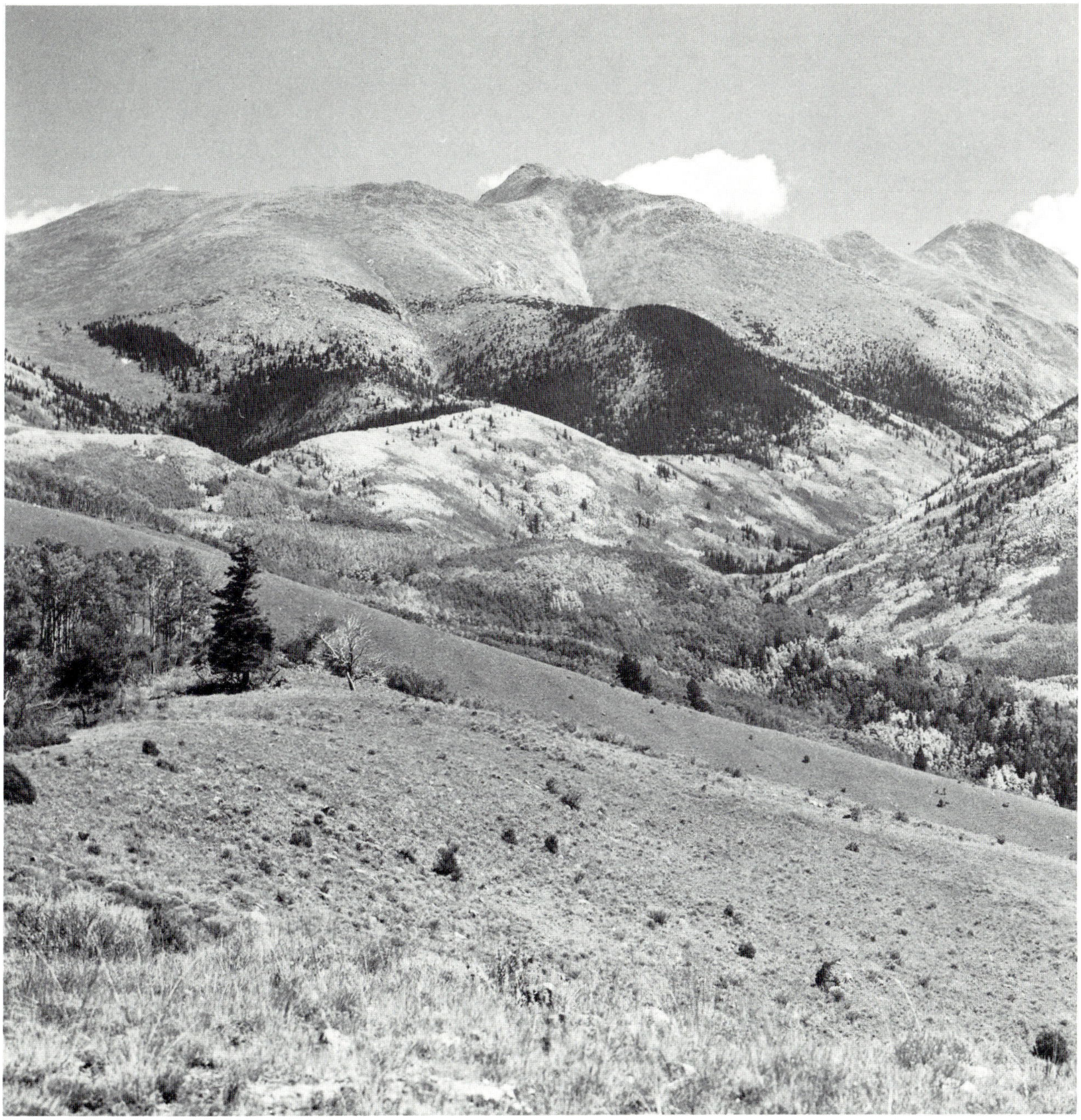

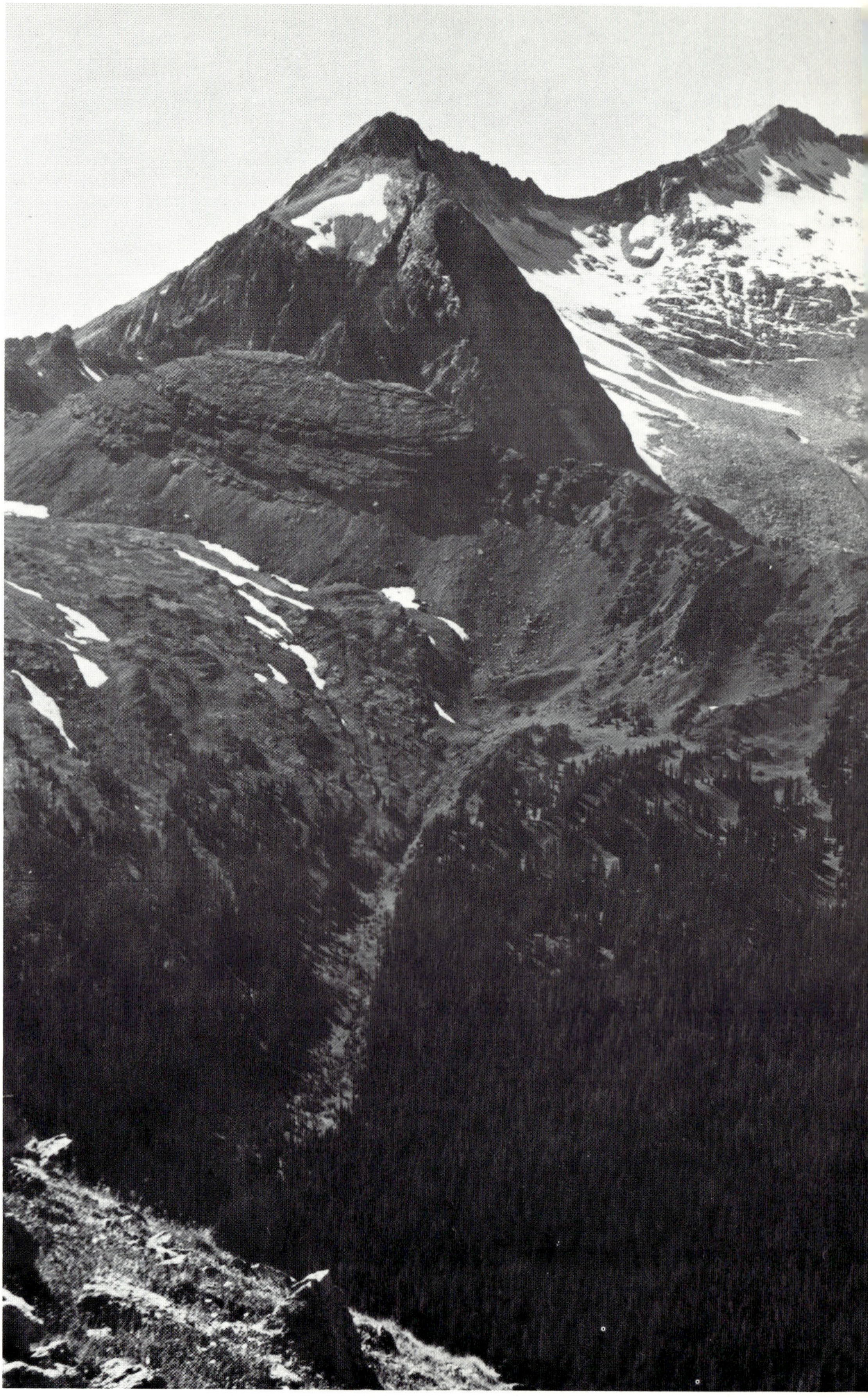

SNOWMASS PEAK (Elk Range) 14,092 feet
32nd highest

Snowmass is well known in Colorado legend and mining history. Often veiled in clouds, it was called "the Cold Woman" by Indians, who believed the mountain was the source of local weather. Some pioneers called the peak "White House Peak," but it was most familiar from the first as Snowmass because of the giant snowfield, visible even in late summer, on the east face. Hayden, who called it Snowmass, said the snowfield was fully five square miles.

Although U.S. Geologist F. V. Hayden was impressed by the mountain, he did not believe it was a Fourteener, calculating that it was 30 feet too short. Snowmass wasn't reassessed until recent years and was among one of the last peaks to be judged a Fourteener. This is another Fourteener that is sometimes referred to as "mountain" instead of "peak."

Because of its beautiful and gentle slopes, Snowmass is a favorite of alpinists, particularly beginners. Favorite base camps are Snowmass and Geneva Lakes; Snowmass is often climbed along with Hagerman in the same day. The only hazard in climbing Snowmass is loose rock near the summit. Professor Whitney said an industrious man with a crowbar could reduce the height of the peak by 100 to 200 feet.

CAPITOL PEAK (Elk Range) 14,130 feet
30th highest

Negotiating the last lap up Capitol Peak requires such risky maneuvering that the mountain is considered one of Colorado's most difficult climbs. Just before reaching the summit, one must traverse a 100-foot knife edge that looks down on a 1,500-foot drop. The intrepid Phil Schmuck has negotiated this edge by sliding forward on the seat of his pants, which is the safest approach. Fortunately for climbers, the knife edge is made of hard granite, rather than the loose sandstone which characterizes so much of the Elk Range to which Capitol belongs. There are a half dozen interesting approaches to the notorious knife edge. The most popular of these routes is via Pierre Basin, an area of lakes above timberline. Pierre and other basins on the mountain were formed by primeval glaciers which melted away millenniums ago.

Capitol was probably first climbed by members of the Hayden Survey in the 1870's. However, Percy Hagerman's ascent on August 22, 1909 was widely touted at the time as the first official climb. The Outward Bound School, which generally includes a Capitol climb in its curriculum, has marked a trail up the mountain to the summit ridge. Getting lost on Capitol is not really a problem, however, since it is one of those rare peaks where the summit is almost always in sight.

Early prospectors called this and Snowmass "The Twins," but Hayden named Capitol for its stately appearance. Though Hayden had estimated that the peak was three feet short of being a Fourteener, those three feet were added a short time later by other surveys. It was then considered an even 14,000 feet until recent surveys again reassessed it to the present height.

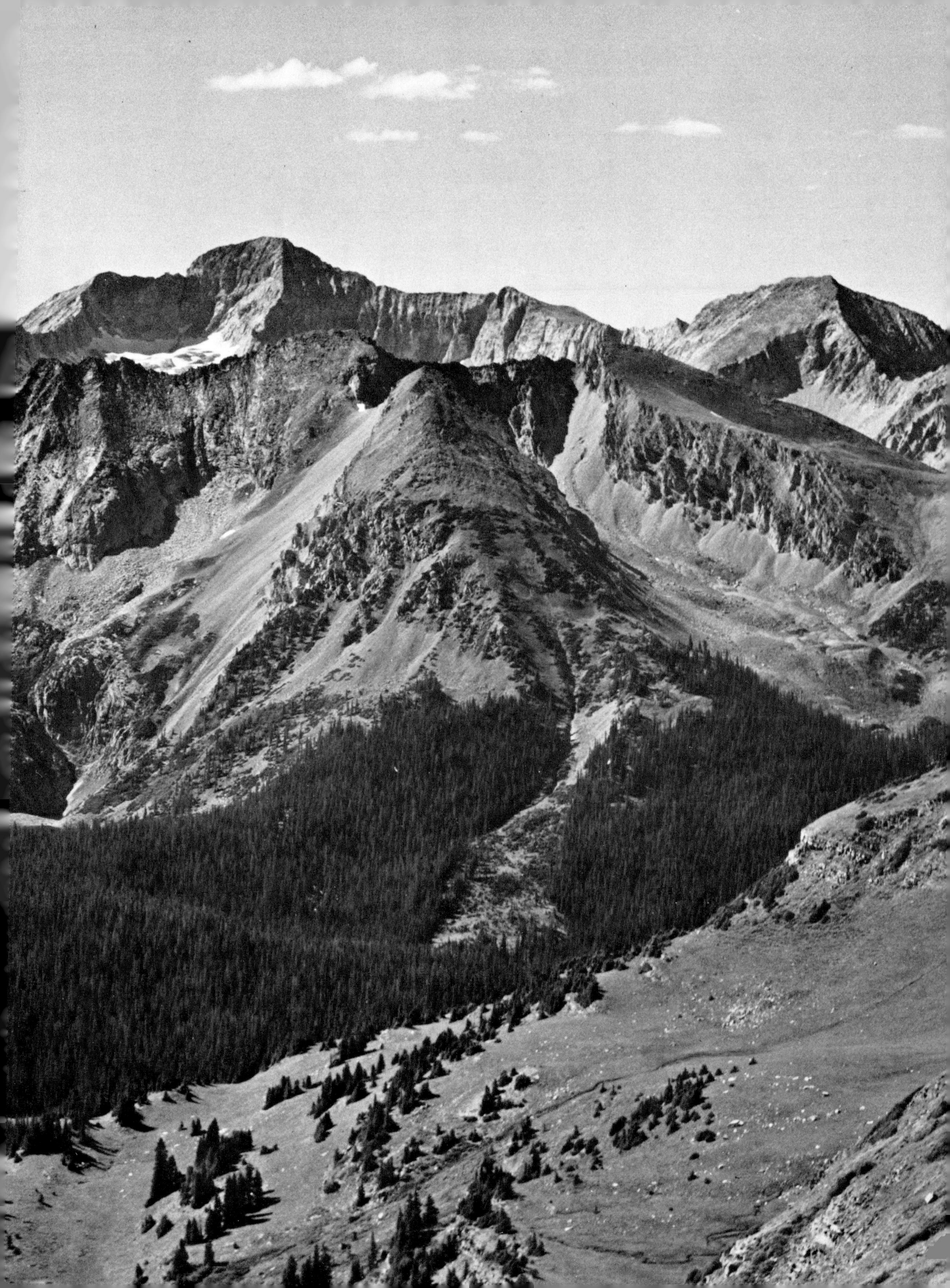

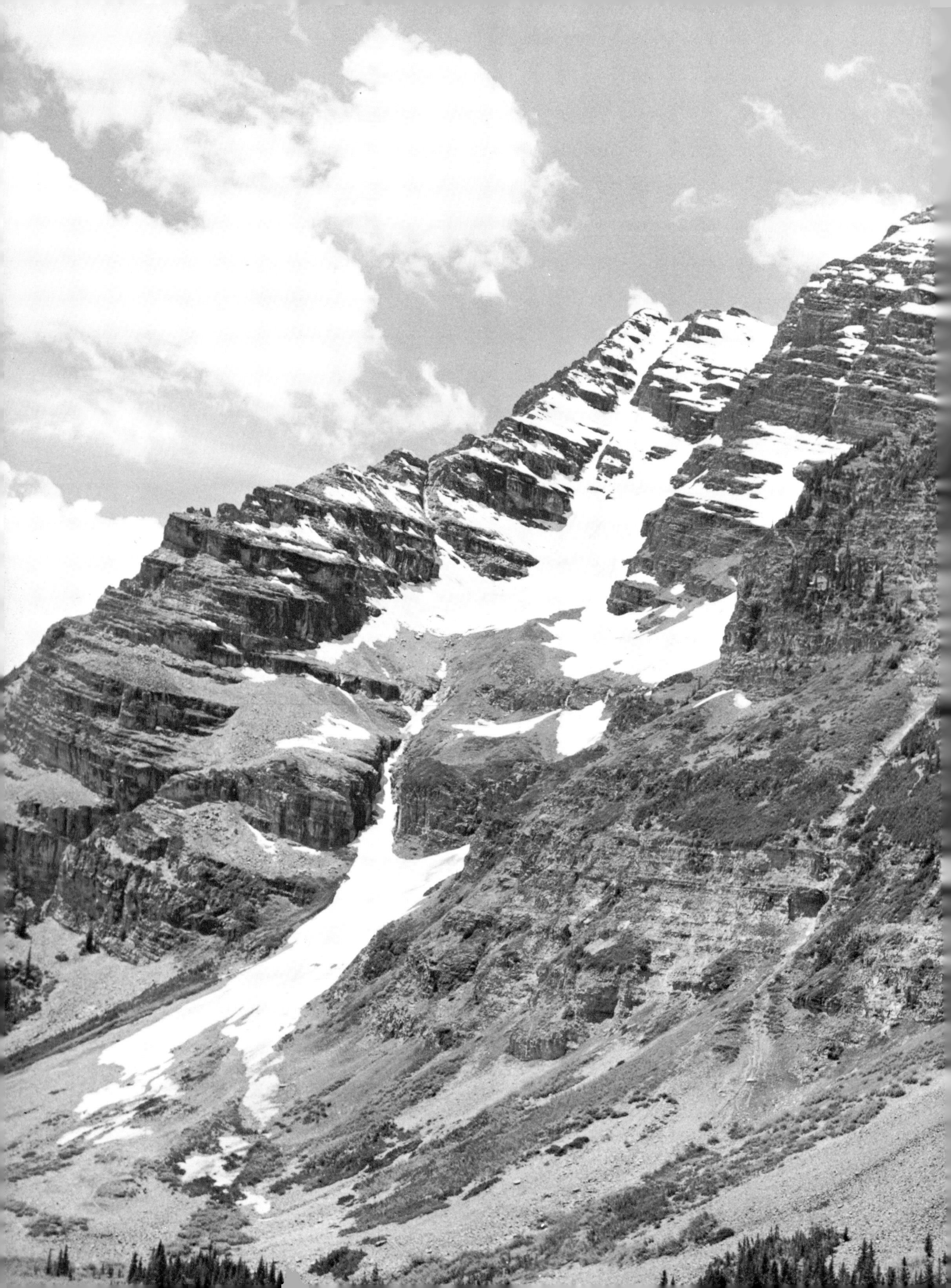

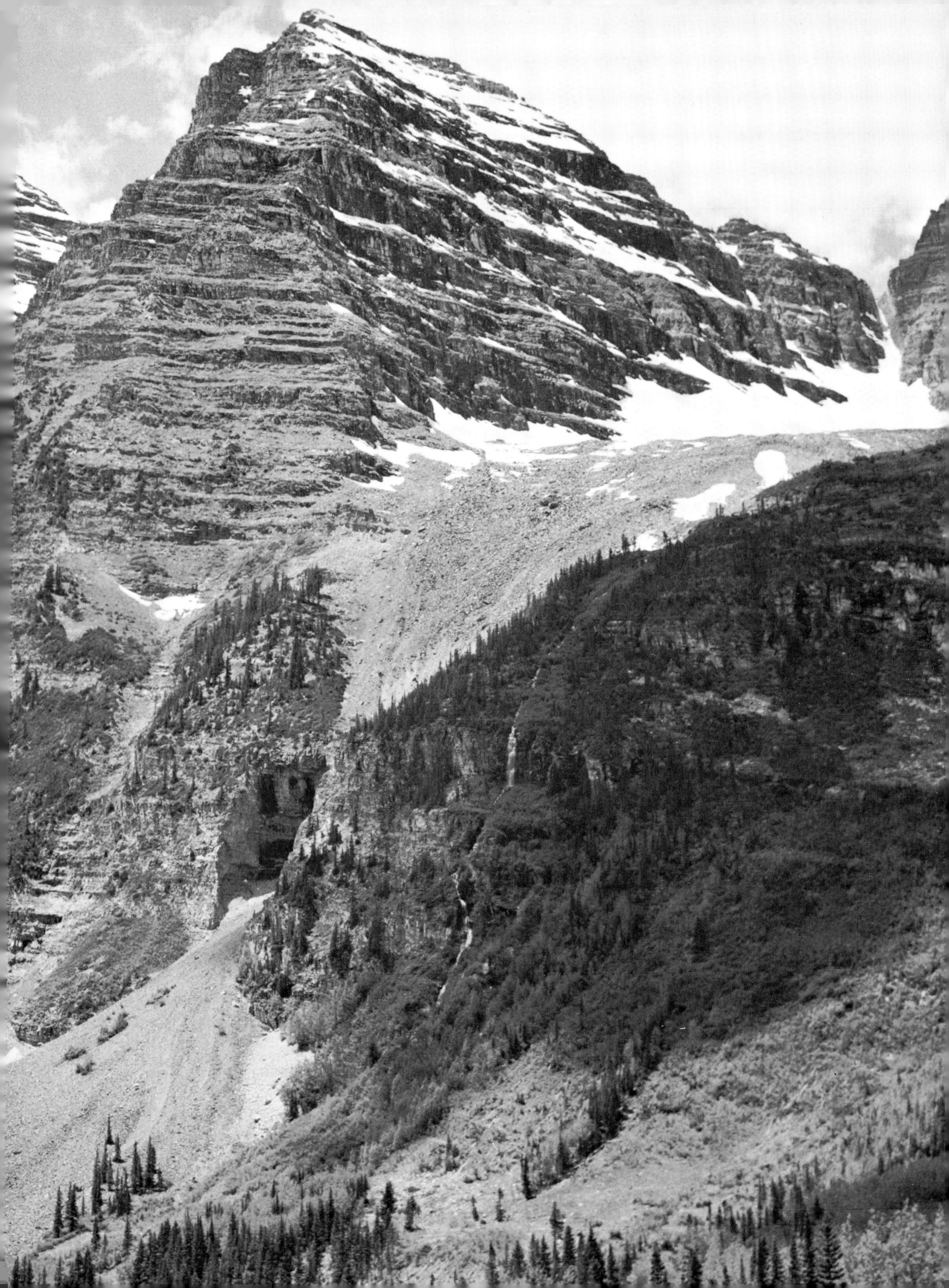

NORTH MAROON PEAK
(Elk Range) 14,014 feet
49th highest

North Maroon Peak and its partner, Maroon Peak, are Colorado's most photographed mountain couple. Sometimes referred to as the Maroon Bells, this pair of mountains is often said to resemble two bells caught in the act of ringing. The reddish, sedimentary mountains display prominent ridges which tilt about 15° from north to south and run almost perfectly parallel along the entire length of mountain face. The basic mountain rock is loose and broken, although the ridges are composed of harder strata and have more successfully resisted erosion.

Early surveyors and mapmakers apparently didn't even bother to separate one peak from the other in their reports. F. V. Hayden and other survey teams simply named the two "Maroon Mountain." Hayden estimated that Maroon or South Maroon Peak was a Fourteener by three feet, but that North Maroon was not a Fourteener at all. It was eventually found to qualify at an even 14,000 feet. A 1959 U.S.G.S. measurement added 14 feet to its official height.

MAROON PEAK or SOUTH MAROON PEAK
(Elk Range) 14,156 feet
25th highest

F. V. Hayden gave future alpinists fair warning back in 1874 when he wrote in his report that the mountain was "nearly, if not quite, inaccessible." Maroon Mountain, as he referred to it, is actually two separate Fourteeners, now known as the Maroon Bells.

Though Maroon Peak is not a great challenge to experienced mountain climbers, the less experienced have sometimes tragically underestimated the challenge. Seven good — not expert — climbers lost their lives within a nine-month period spanning the summer of 1965 and the spring of 1966. Victims included three top scientists from Los Alamos, New Mexico who fell several hundred feet down the mountain. A fourth member of the Los Alamos team also fell but sustained only minor injuries.

The experience of 19-year-old Joe Fullop was even more incredible. He and two fellow Western State University students made the summit of Maroon Peak at sunset on April 24, 1966, but the chilling wind made them decide to start back down immediately. Roped together, the three climbers suddenly plunged 1,000 feet. It is believed that their nearly frozen, insensitive fingers gave out and caused them to plummet from the mountain. Young Fullop, anchor man on the 125-foot rope, fell the furthest and was knocked unconscious. His companions were killed, but by some miracle Fullop managed to survive.

The many tragedies that occurred on the mountain within such a short period of time gave newspapermen and phrasemakers a field day for word-weaving. Maroon Peak and North Maroon were called "the mountains of frozen death." One sardonic scribe claimed the victims were those "for whom the Bells tolled."

Most of the victims were claimed by Maroon Peak, or South Maroon Peak as it is also known. However, most of the falls occurred on the way down and between the peaks, near Maroon Lake.

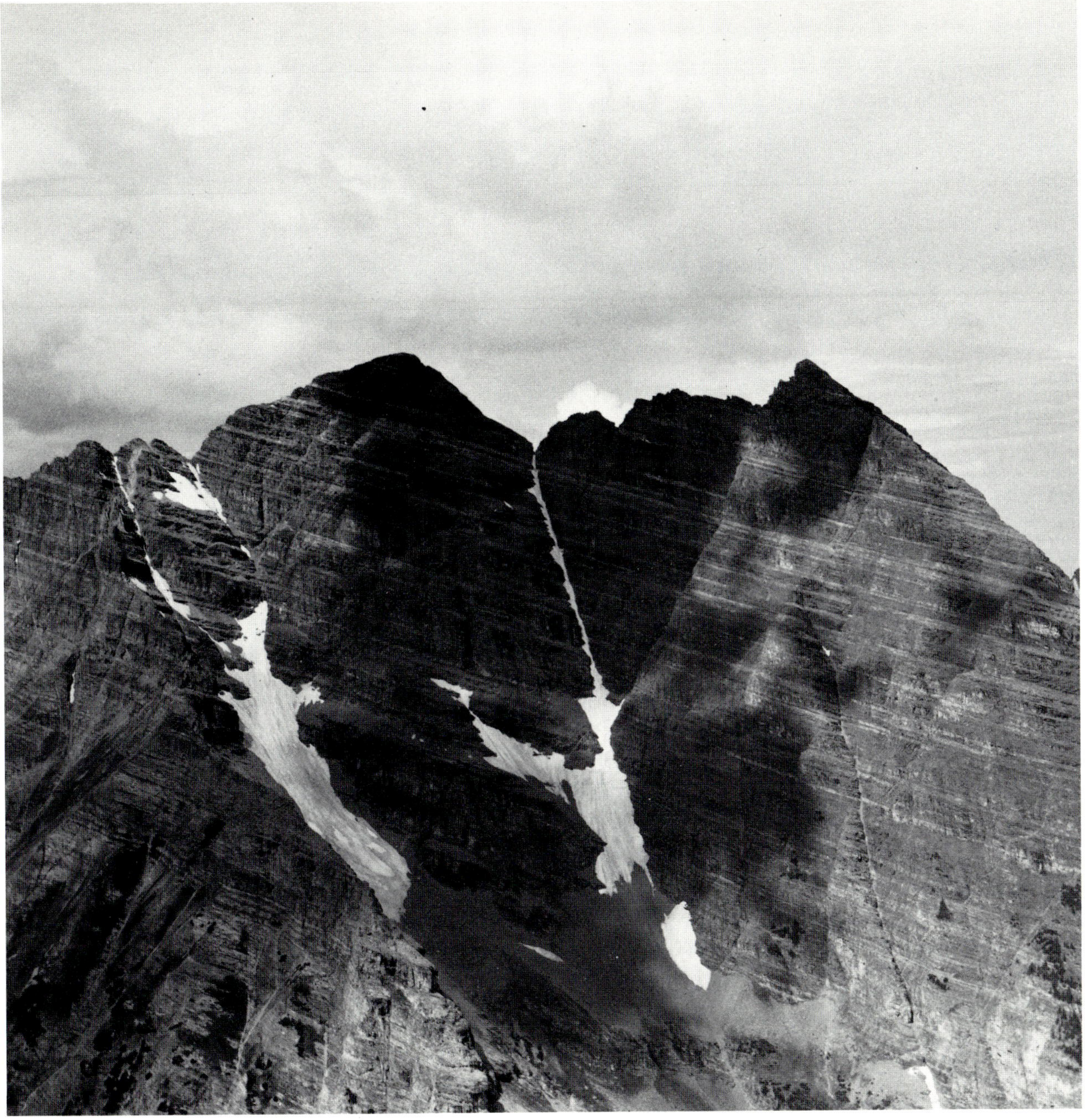

PYRAMID PEAK (Elk Range) 14,018 feet
46th highest

When G. M. Wheeler dubbed this one of Colorado's "most spectacular" mountain peaks, he also warned would-be climbers that it was one of the state's toughest climbs. Neither the Wheeler nor the Hayden Survey team was able to reach the top of the peak. One member of the Wheeler party did come within 200 feet of the summit, but decided that to continue was "mere recklessness."

The climb is still considered one of the most hazardous in the state, not so much for its steep slopes, but because they are composed of loose sandstone which makes footing difficult.

It is possible that the peak was climbed in the 1880's or 1890's, but the first recorded climb was made, amid much hoopla, by Percy Hagerman in 1909.

Although Wheeler was greatly impressed with the mountain, he calculated that it was only 13,885 feet high. Other early surveys listed the peak as an even 14,000 feet, and so it remained until recent years.

Hayden, who surveyed the area at about the same time Wheeler did, originally called the peak "Black Pyramid." The name has subsequently been shortened to Pyramid.

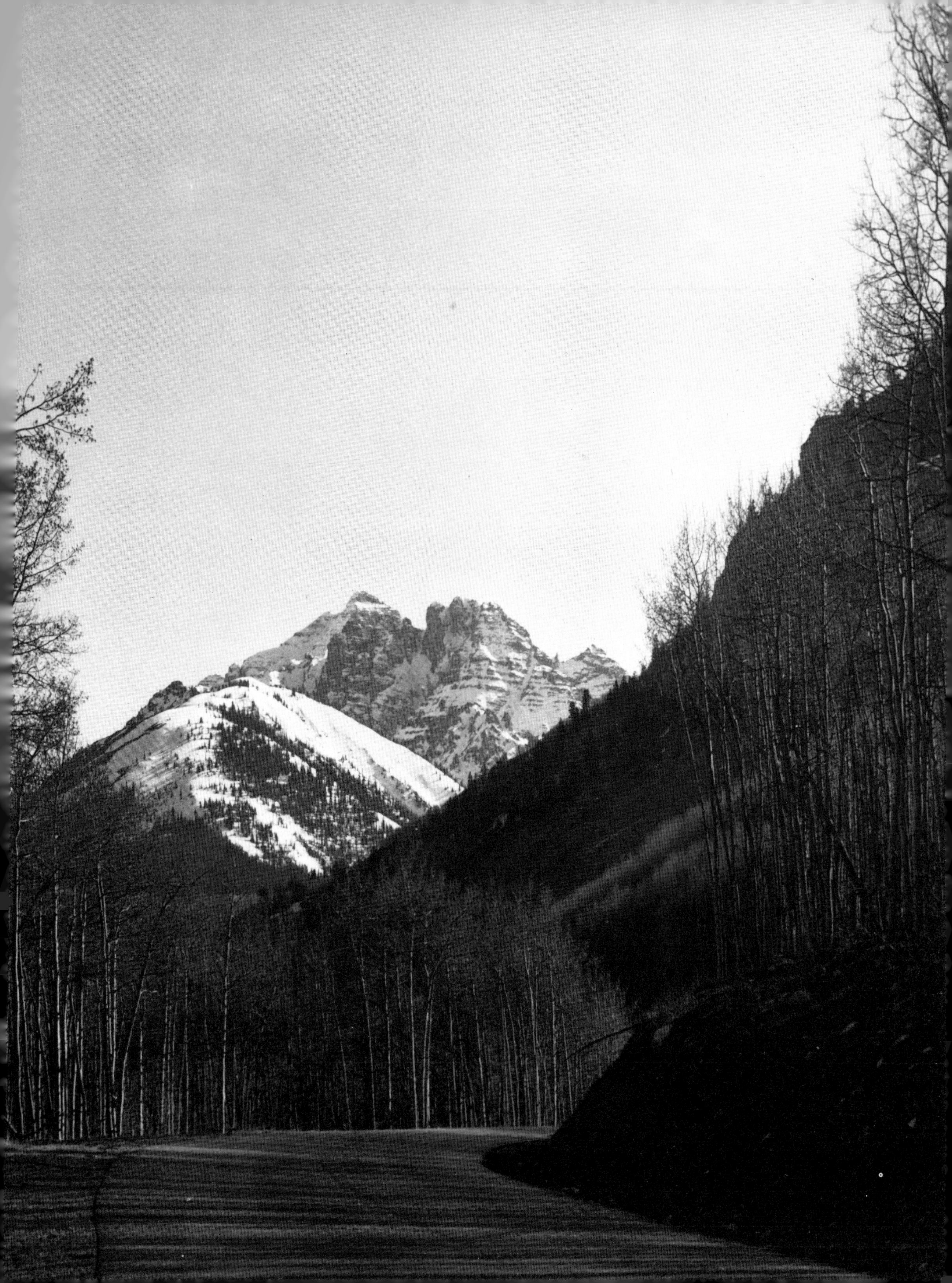

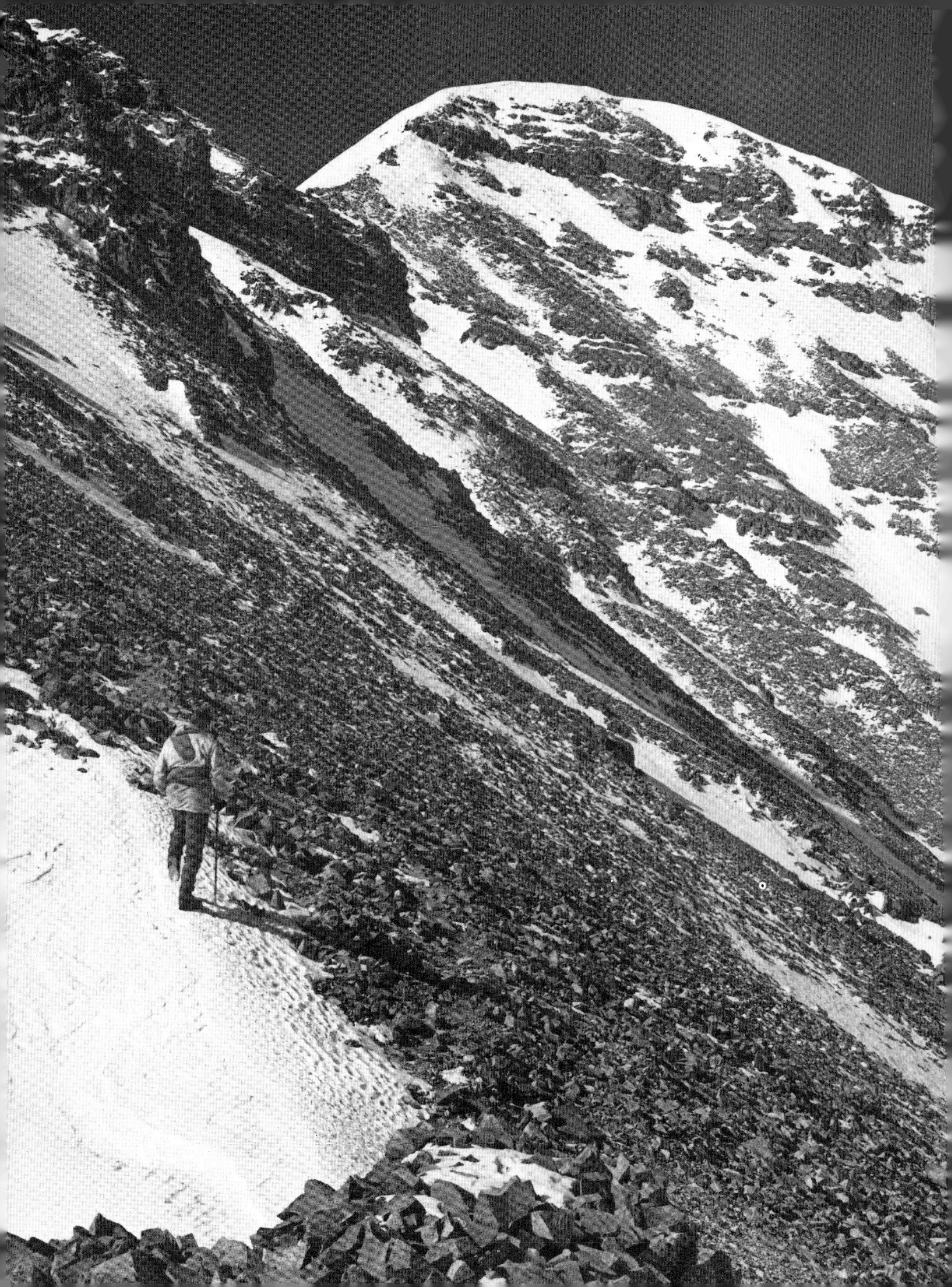

CASTLE PEAK (Elk Range) 14,265 feet
12th highest

The summit of this peak is the highest point in the Elk Range. Although it is not as well known as some of its more illustrious (and shorter) neighbors, it is one of the loveliest. Its color is dominated by purples, and austere ridges lend it the impressive appearance of a castle.

It is called an "interesting" climb, although not as difficult as many others in the Elk Range. Castle Peak was named and first climbed by the Hayden Survey in the mid-1870's. A detailed description of Henry Gannett's climb up the mountain in 1902 appeared in *Everybody's Magazine.*

SAN LUIS PEAK (San Juan Range) 14,014 feet
50th highest

This peak is separated from Stewart Peak by a long ridge and is usually climbed with its neighbor.

San Luis was believed named for the valley below, which feeds on the water drained into the Rio Grande from the melting snow on its slopes.

Unlike its neighbor, Stewart Peak, which has gained 30 feet in recent years, San Luis has lost about 130 feet; but it is still a Fourteener, though at one earlier time San Luis was considered to be only 13,146 feet.

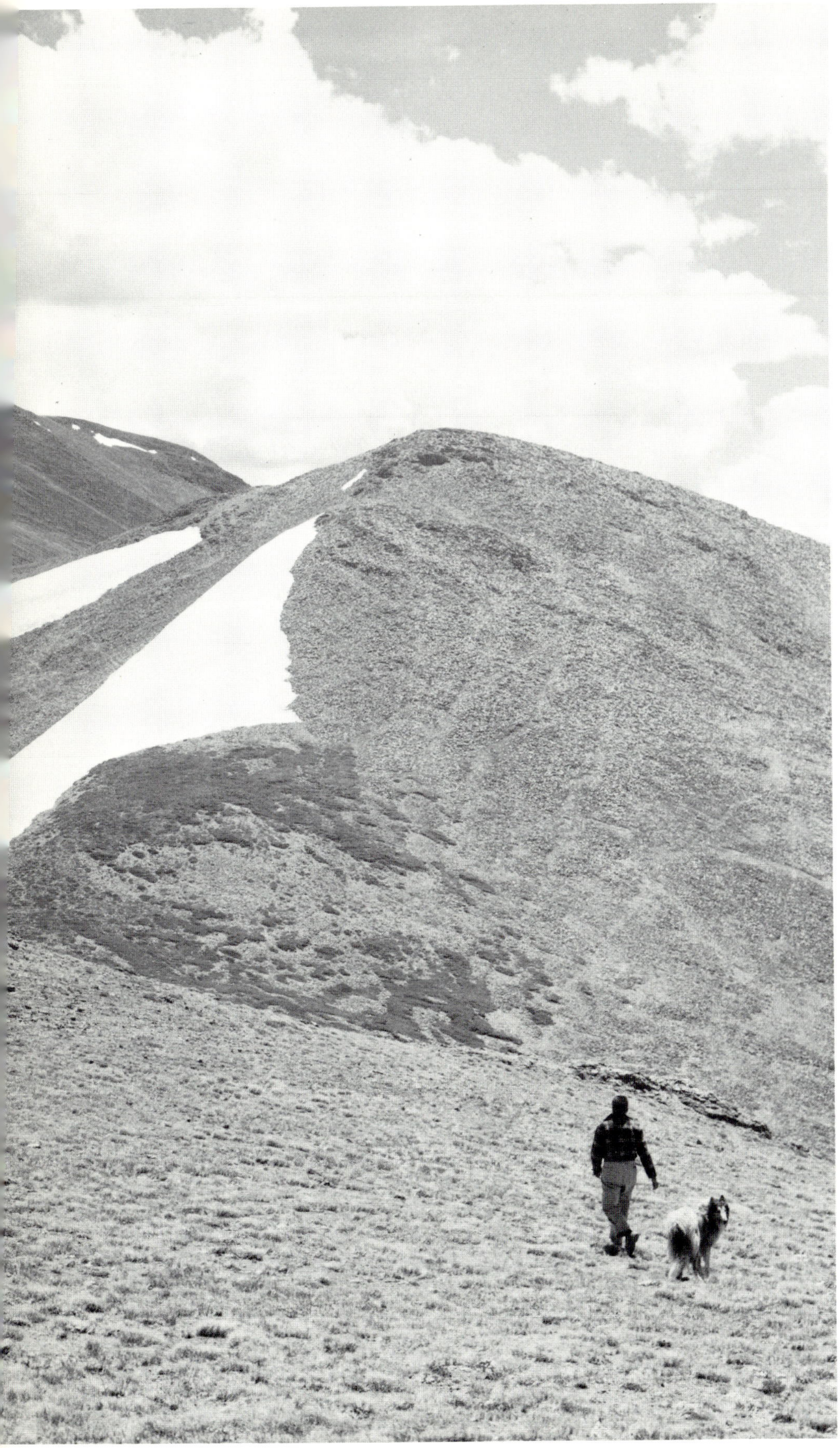

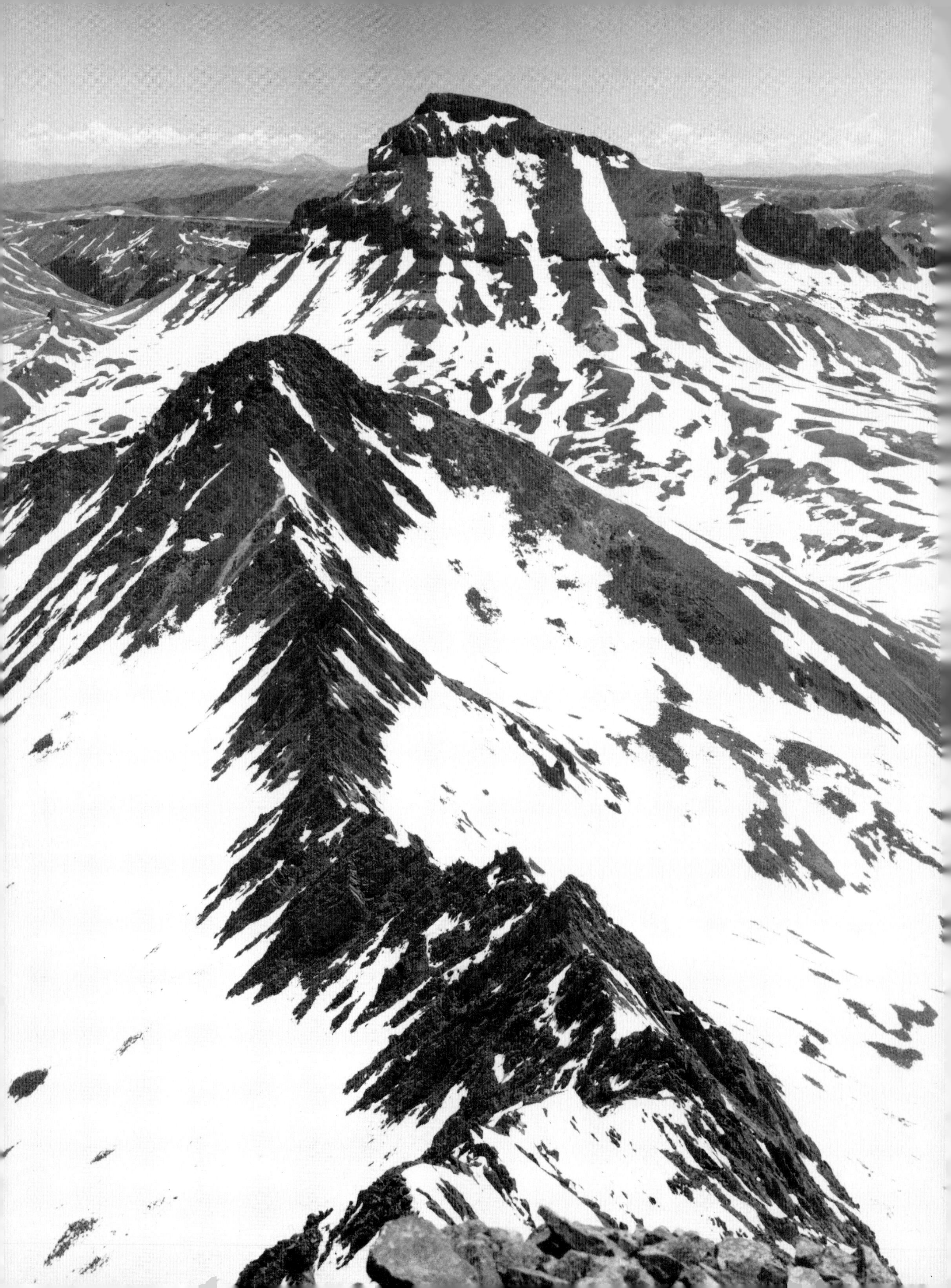

UNCOMPAHGRE PEAK
(San Juan Range) 14,309 feet
6th highest

Uncompahgre was first climbed by Franklin Rhoda and A. D. Wilson of the 1874 Hayden Survey during a violent thunderstorm. Theirs was also the first taste of a spectacular San Juan phenomenon that many were to experience in the years that followed. Lightning striking nearby made their hair stand on end like stiff porcupine quills, sharp rocks sizzled like frying bacon, and their transit sang like a telegraph key. As Rhoda wrote, "We experienced, for the first time in the season, the electrical phenomena which later interfered so much with the topographical work . . . Our experience on all the peaks was very similar."

In 1881, when this peak was considered by many to be the state's highest, more than 40 mining men climbed the peak and placed a flag at half mast on the summit as a memorial to the recent death of President James A. Garfield. Although Uncompahgre wasn't actually the highest mountain in Colorado, that was undoubtedly the "highest tribute" given the late president. These mining men also encountered lightning. Though a bolt struck in the middle of the group and injured a man, they kept the flag flying.

Few if any dispute that the mountain was named for the Uncompahgre River, but there is some difference over the literal meaning of the word "uncompahgre." Some interpret it to be a distortion of the Ute word "ancapogari" meaning "red lake." The Indians presumably gave the Uncompahgre River this name because of the spring of reddish water, hot and disagreeable to the taste, near the river's source. Others believe the original Ute word would be translated "hot" (unca) "water" (pah) "spring" (gre). As early as 1853 the peak was called Uncompahgre by Lieutenant E. G. Beckwith, official recorder for the Gunnison Expedition. However, the mountain was also called "Mount Chauvenet" for a professor of astronomy at Washington University in St. Louis, and mining men at times referred to the peak as "The Leaning Tower" and "Capitol Mountain."

Uncompahgre is an interesting climbing mountain. There are some difficult trails, but its gentle southern slopes are considered an easy "walk-up." A miner's cabin, in fact, has been found as high as 12,500 feet on this peak. However, the formidable northern face of the mountain, which is a sheer cliff of several thousand feet of crumbling, rotting stone, has never been climbed.

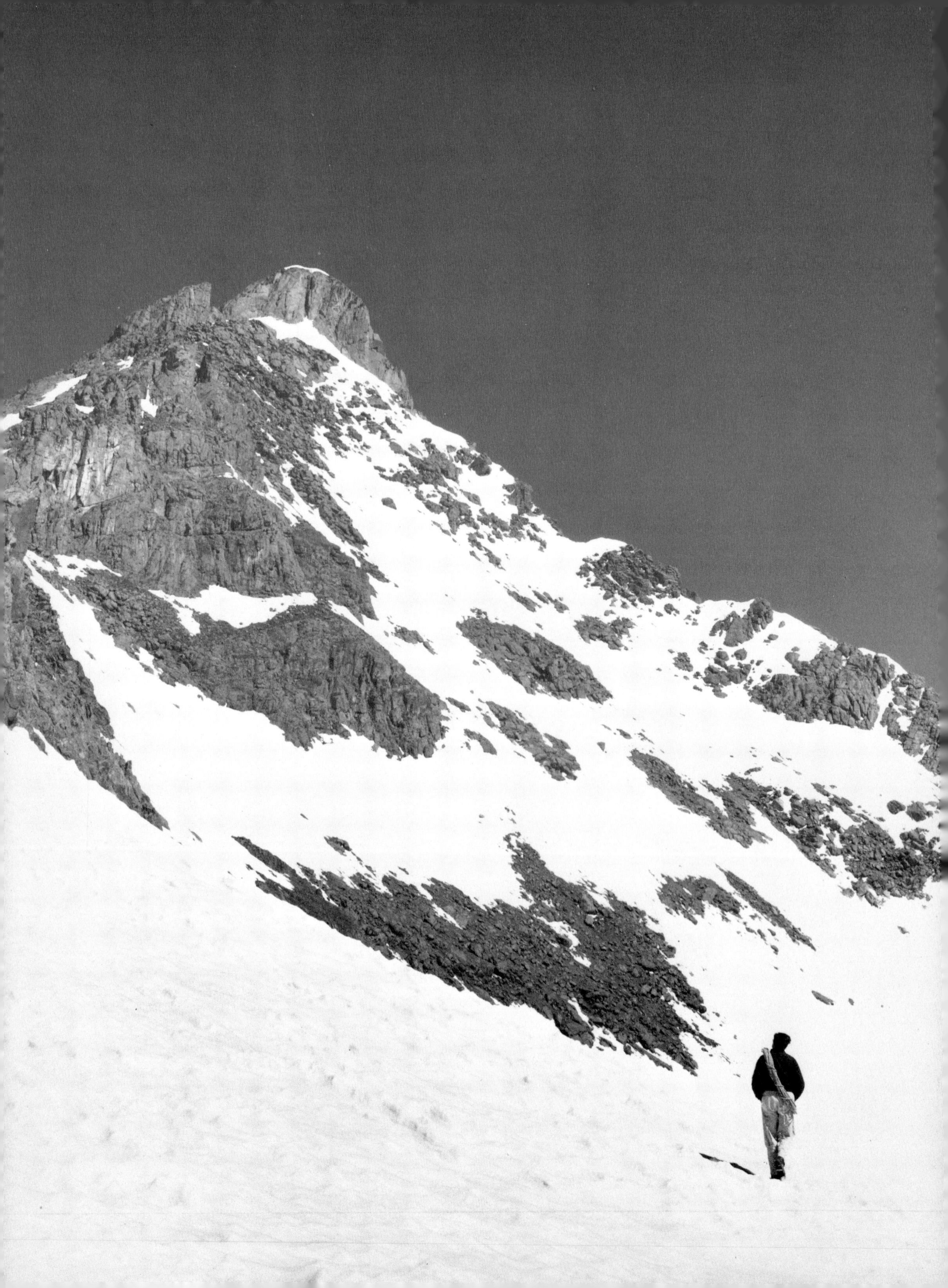

WETTERHORN PEAK
(San Juan Range) 14,017 feet
48th highest

It is believed Wetterhorn was named by the Wheeler Survey party in 1874 for the famed Swiss peak which some believe it resembles.

Not considered a particularly difficult climb, it was probably climbed initially by prospectors. There has been much mining around the mountain.

SUNSHINE PEAK (San Juan Range) 14,001 feet
53th highest

Some experts have long suspected Sunshine to be less lofty than the records indicated. A few believed the peak would lose its Fourteener status with more accurate measurements. Sure enough, the latest U.S.G.S. calculation dimmed Sunshine's height 17 feet, but left it in the club — barely.

The summit of Sunshine is less than two miles from the summit of Redcloud, and both are often climbed in one day. Sunshine was called "Niagara Peak" or "Sherman Mountain" in the past. The Geological Survey gave the peak its present name in 1904 or 1905, without offering any reason for their choice. In 1874, Hayden Survey members Franklin Rhoda and A. D. Wilson had simply designated the peak as "Station 12" and had used it as one of their headquarters. Rhoda and Wilson were the first officially to climb Sunshine; it was this climb that Rhoda wrote about in detailing the strange happenings first experienced less spectacularly on Uncompahgre Peak.

"We had scarcely got started to work when we both began to feel a peculiar tickling sensation along the roots of our hair, just at the edge of our hats, caused by the electricity in the air. By holding up our hands above our heads a tickling sound was produced, which was still louder if we held a hammer or other instrument in our hand . . . a peculiar sound almost exactly like that produced by the frying of bacon. This latter phenomenon, when continued for any length of time, becomes highly monotonous and disagreeable. We felt that we could not stop, though the frying of our hair became louder and more disagreeable, for certain parts of the drainage of this region could not be seen from any other peak, and we did not want to ascend this one a second time.

"As the force of the electricity increased, and the rate of increase became greater and greater, the instrument on the tripod began to click like a telegraph-machine when it is made to work rapidly; at the same time we noticed that the pencils in our fingers made a similar but finer sound whenever we let them lie back so as to touch the flesh of the hand between the thumb and forefinger. This sound is at first nothing but a continuous series of clicks, distinctly separable one from the other, but the intervals becoming less and less, till finally a musical sound results. The effect on our hair became more and more marked. When we raised our hats our hair stood on end, the sharp points of the hundred of stones about us each emitted a continuous sound, while the instrument outsang everything else. The points of the angular stones being of different degrees of sharpness, each produced a sound peculiar to itself . . .As I took the barometer out of its leather case, and held it vertically, a terrible humming commenced from the brass ring at the end, and increased in loudness so rapidly that I considered it best to crawl hastily down the side of the peak to a point a few feet below the top, where, by lying low between the rocks, I could return the instrument to its case with comparative safety. At the same time Wilson was driven from his instrument, and we both crouched down among the rocks to await the relief to be given by the next stroke [of lightning], which, for aught we knew, might strike the instrument which now stood alone on the summit. At this time it was producing a terrible humming, which, with the noises emitted by the thousands of angular blocks of stone, and the sounds produced by our hair, made such a din that we could scarcely think. The fast-increasing electricity was suddenly discharged, as we had anticipated, by another stroke of lightning, which, luckily for us, struck a point some distance away. The instant he felt the relief, Wilson made a sudden dash for the instrument, on his hands and knees, seized the legs of the tripod, and flinging the instrument over his shoulder dashed back. Although all this occupied only a few seconds, the tension was so great that he received a strong electric shock, accompanied by a pain as if a sharp-pointed instrument had pierced his shoulder, where the tripod came in contact with it. In his haste he dropped the small brass cap which protected the object-glass of the telescope; but, as the excitement and danger had now grown so great, he did not trouble himself to go back after it, and it still remains there in place of the monument we could not build to testify to the strange experiences on this our station 12. We started as fast as we could walk over the loose rock, down the southeast side of the peak, but had scarcely got more than 30 feet from the top when it was struck. We had only just missed it, and felt thankful for our narrow escape."

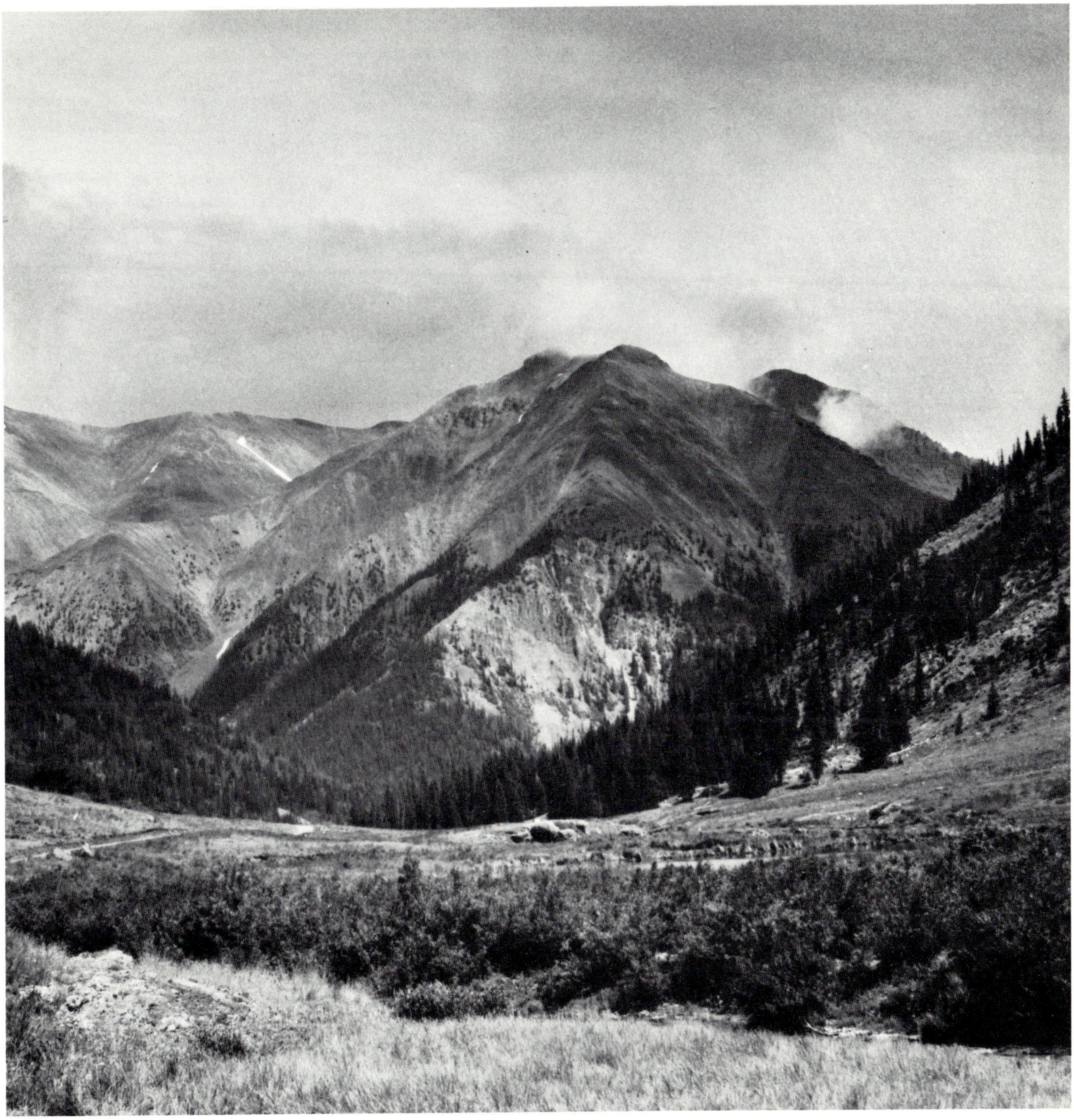

REDCLOUD PEAK (San Juan Range) 14,034 feet
45th highest

The U.S. Geological Survey is currently taking another look at this peak and its neighbor, Sunshine, to see if they have been overrated. Redcloud might lose a few feet but is expected to stay in the Fourteener fraternity.

The peak was earlier known as "Red Mountain" and "Jones Peak," but was officially named Redcloud in 1874 for its dull red appearance and its high ridges that are said to resemble clouds.

Because of its isolated location, this is one of Colorado's least-seen mountains, although also one of the loveliest. Redcloud is considered an easy climb, and most climbers "do" Redcloud and Sunshine in one day.

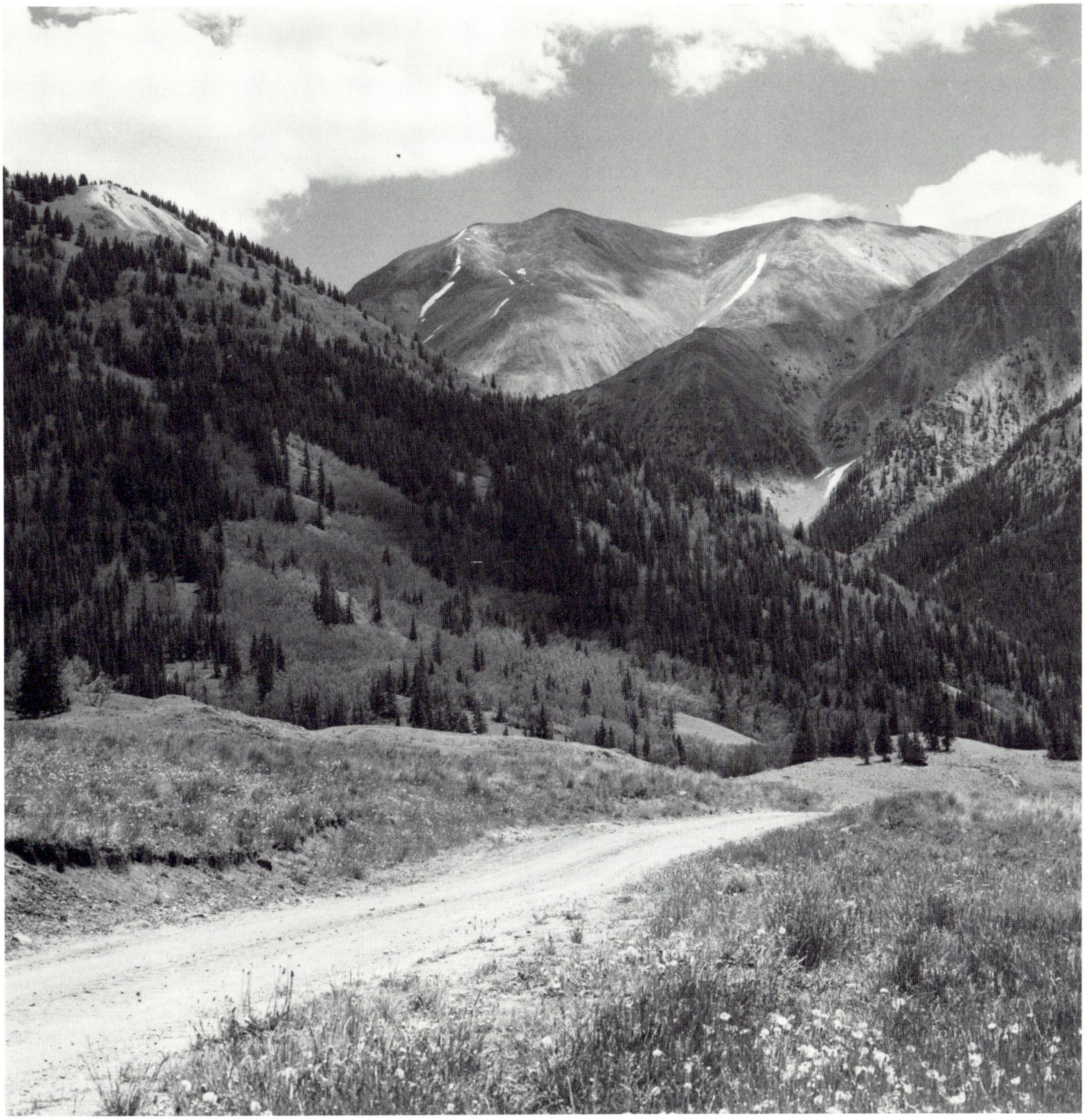

HANDIES PEAK (San Juan Range) 14,048 feet
40th highest

Early Forest Service maps referred to this peak as "Tobasco," but Hayden noted in 1874 that the peak was already known as Handies. Early historian H. H. Bancroft reported that the mountain's namesake was a pioneer who became a prominent resident of the region.

Handies, Redcloud, and Sunshine are near neighbors, and all three were climbed by Rhoda and Wilson of the Hayden Survey in 1875. No real challenge, Handies was also managed easily by early prospectors, who dug and worked prospect holes near its summit. Today's "prospectors" may not be rewarded by the discovery of silver and gold, but one of Colorado's richest views can be enjoyed from Handies' summit.

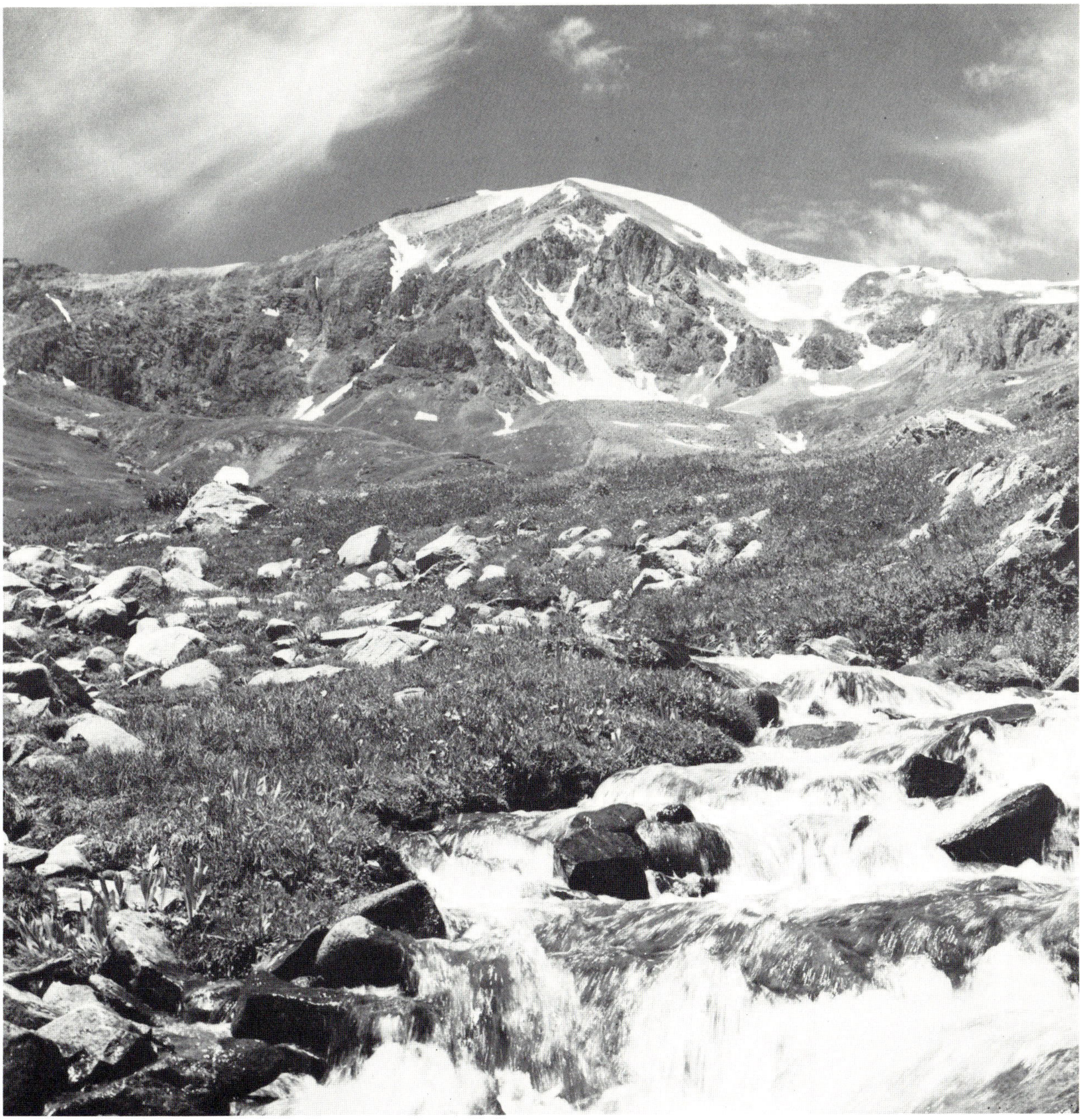

MOUNT EOLUS (San Juan Range) 14,084 feet
34th highest

Mount Eolus is in the Needles Mountains, sometimes popularly referred to as Needles Range, which is one of the most spectacular yet rarely seen ranges in Colorado, located in one of the least accessible areas of the state.

Hayden, after viewing Eolus and its neighbors through thick clouds and strong winds in 1874, is believed to have named this peak for Aeolus, the Greek god of winds.

The Needles Mountains were probably first climbed by early prospectors. Today most alpinists use the old mining camp and stage stop of Needleton as a base camp. Eolus is one of three Fourteeners in the Needles, and the only one not climbed in the winter of 1966 by Phil Schmuck, Kermith Ross, and Don Monk, who were turned back by strong winds and threatening weather.

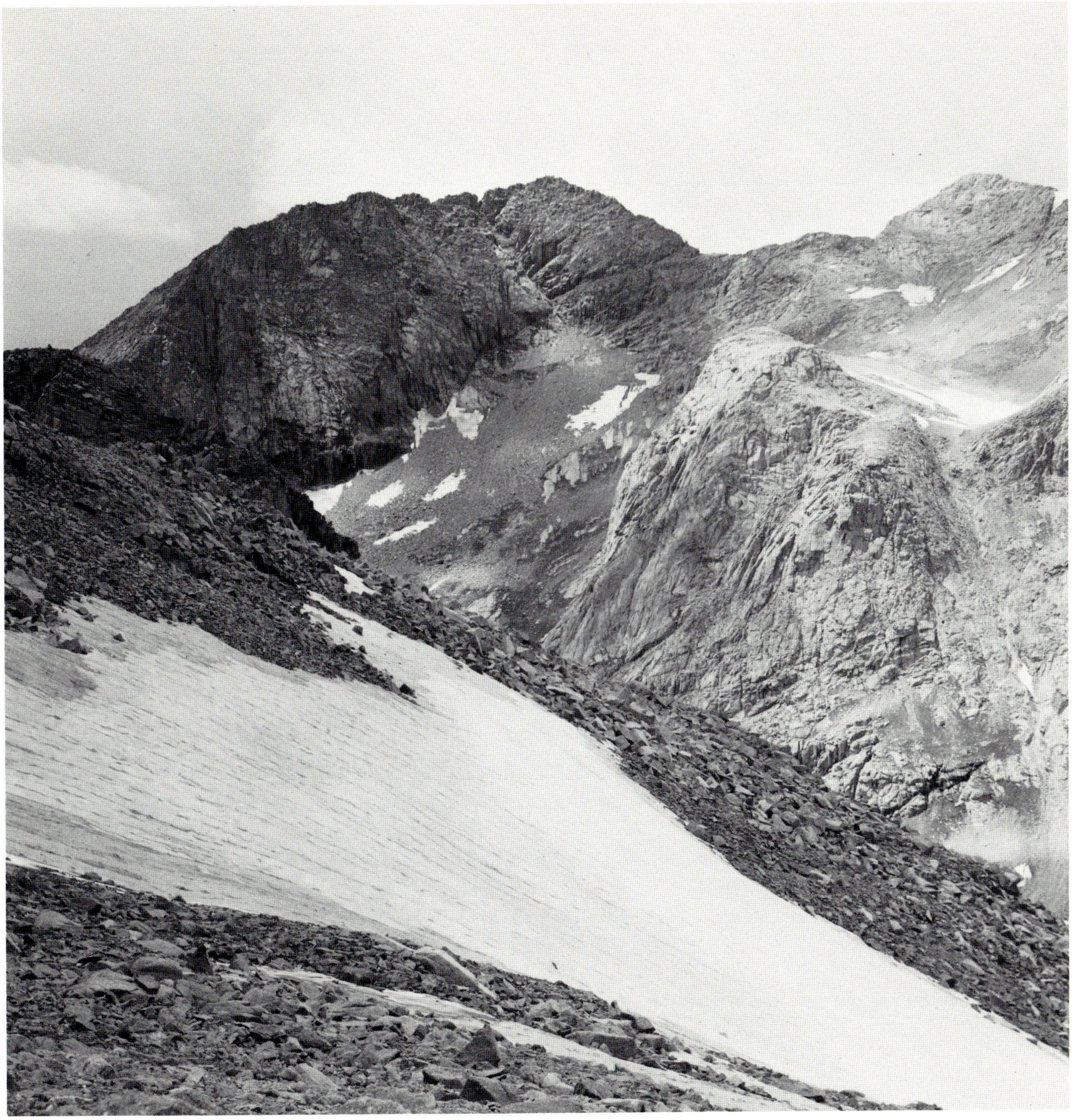

SUNLIGHT PEAK (San Juan Range) 14,059 feet
39th highest

Sunlight is flanked by Mount Eolus and Windom Peak, the other Fourteeners which make up the Needles' rugged mountain silhouette. The cigar-shaped dome of Sunlight adds to its climbing challenge, particularly in winter when its narrow ridges are covered with ice and snow. The first winter climb was accomplished in December 1966 by Phil Schmuck, Don Monk, and Kermith Ross, who were only able to reach the summit via a long and devious route.

Sunlight was named by the Geological Survey when it charted the Needles in 1902.

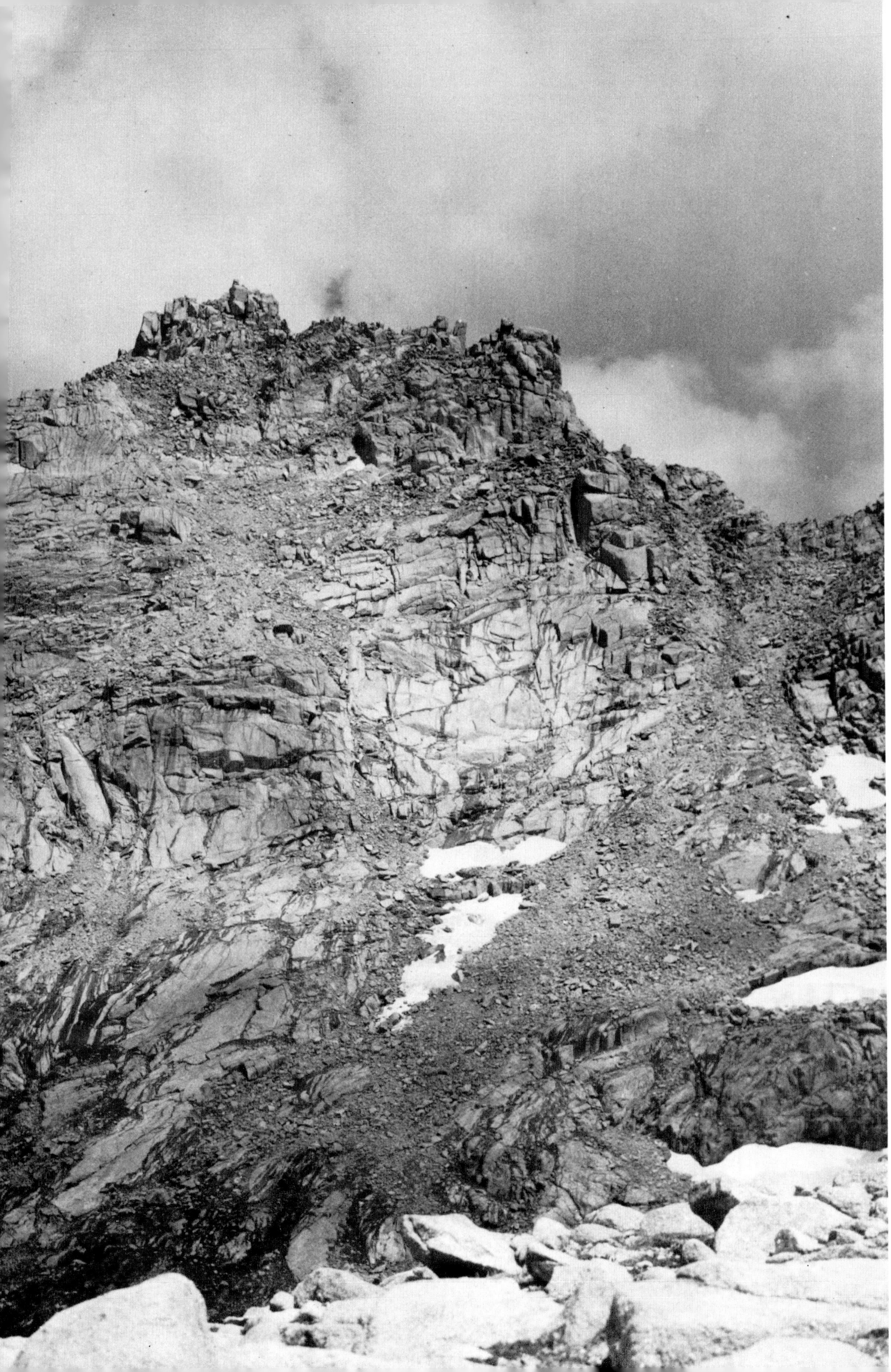

WINDOM PEAK (San Juan Range) 14,087 feet
33rd highest

Many think this peak was poetically named for the strong, ever-present winds on its slopes. Actually, it was named by the Geological Survey in 1902 for William Windom, who, at various times, had served as a U.S. representative, a U.S. senator, and as secretary of the treasury under President James A. Garfield.

During the summer, Windom is not a difficult climb, and the summit affords a magnificent view of the area. In December 1966 Schmuck, Ross, and Monk became the first men to negotiate Windom in winter. The three also made the first winter climb of Sunlight at that time, since — winter or summer — the two peaks are generally climbed together.

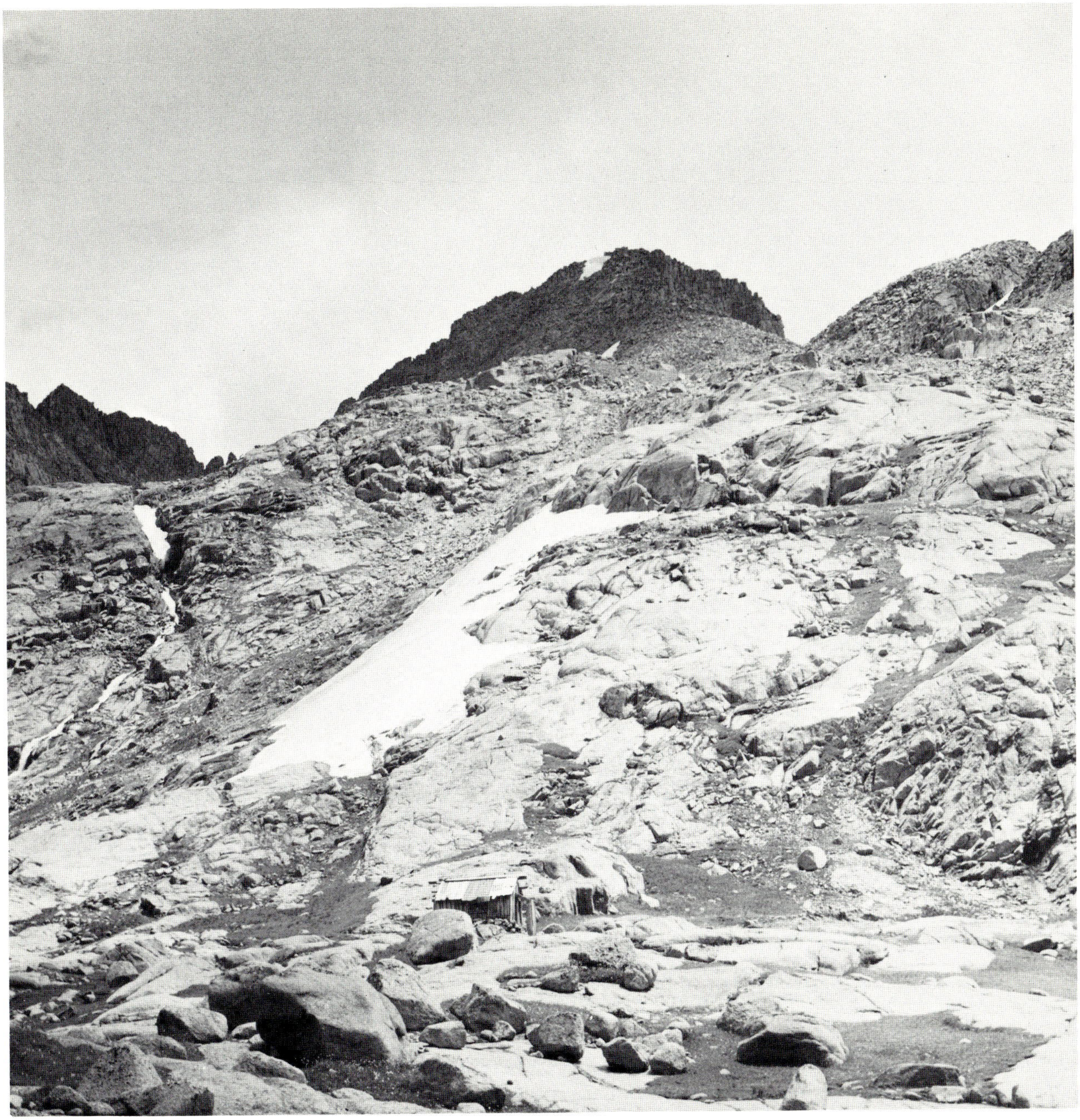

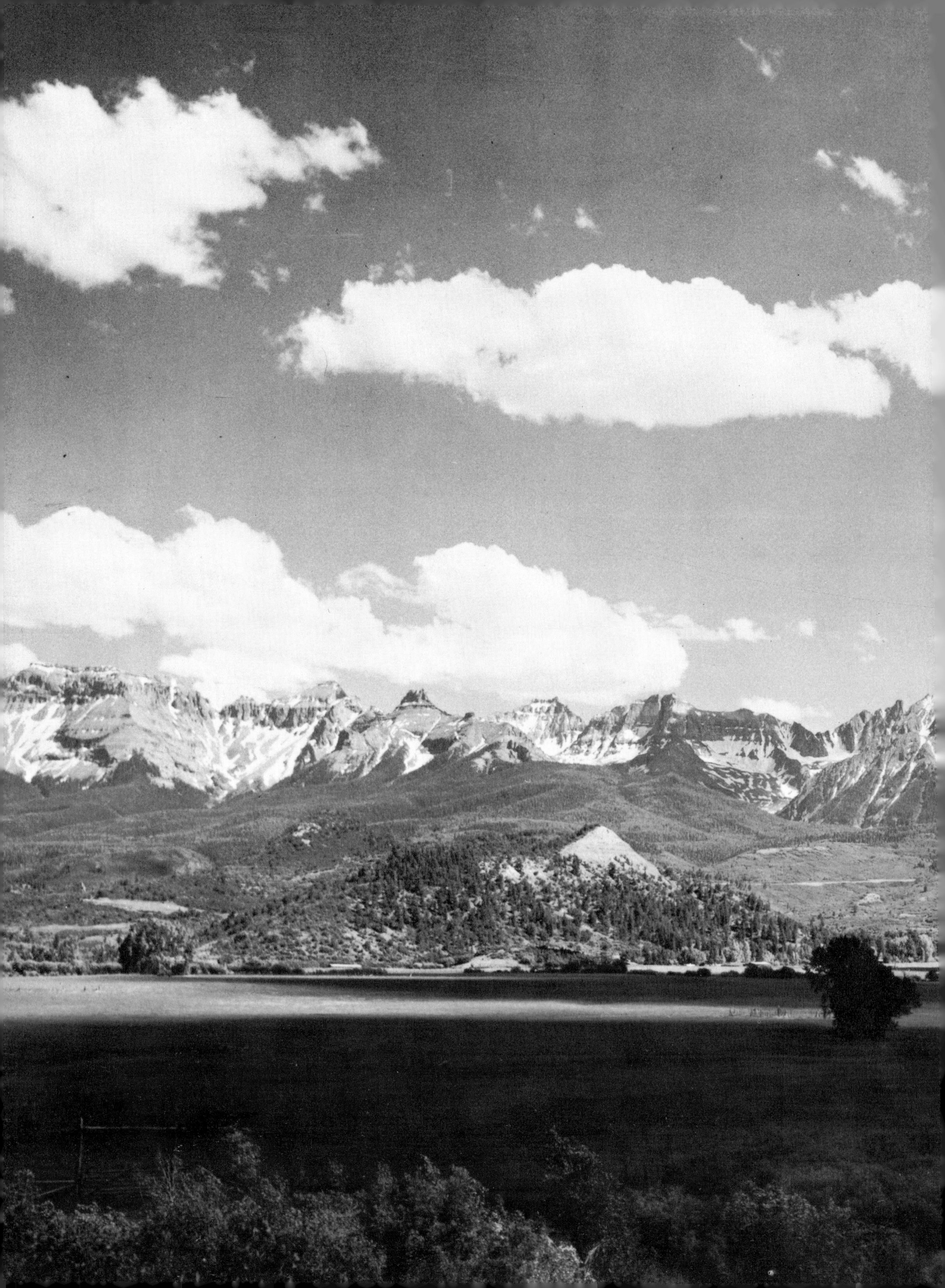

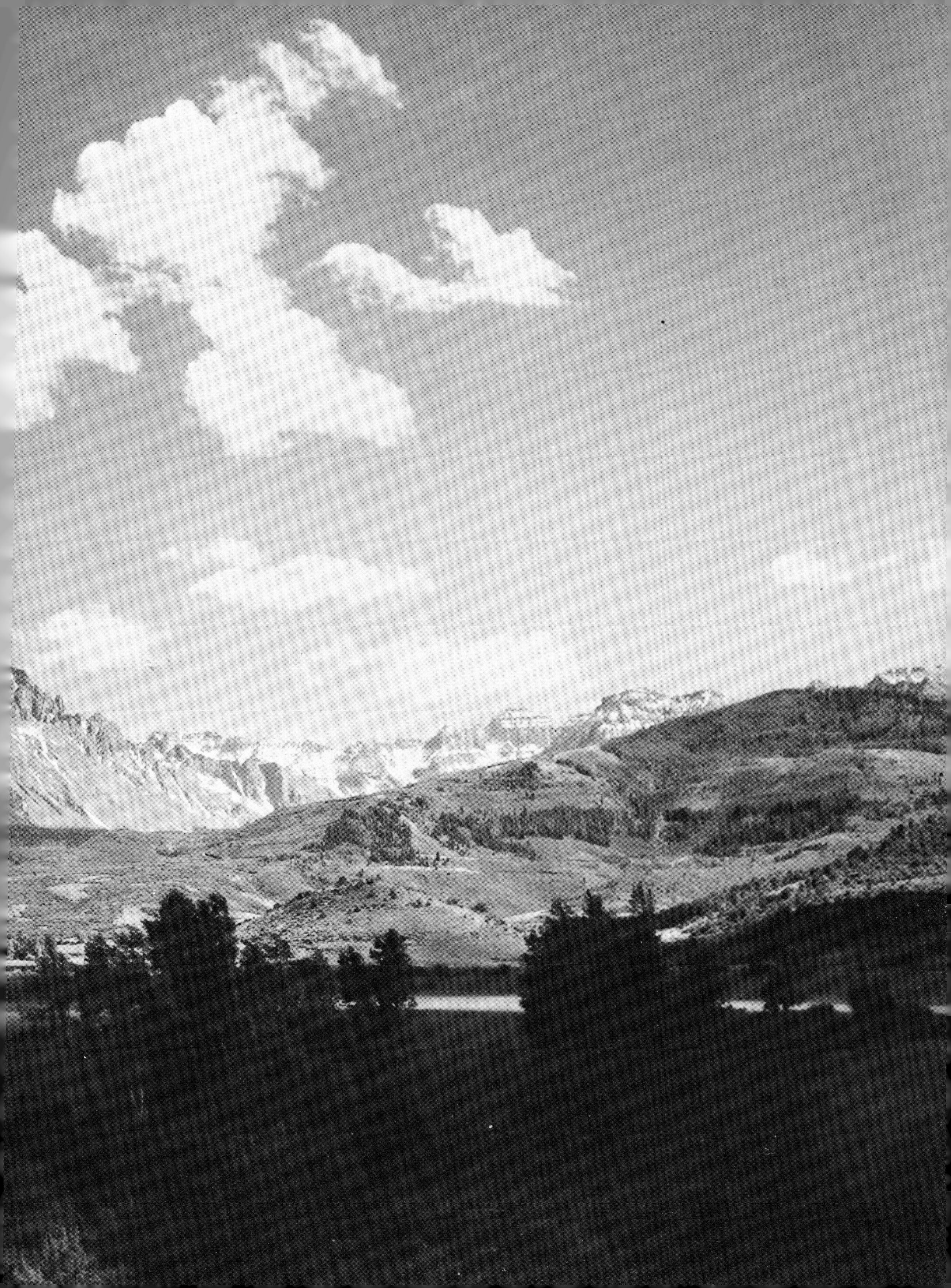

MOUNT SNEFFELS
(San Juan Range) 14,150 feet
28th highest

In 1899 the mines on the slopes of Mount Sneffels had already produced $35 million, making this peak the richest mountain in the nation. Unlike other early sources of gold and silver, Sneffels' ore has never given out, and the mountain is probably still the greatest producer in this country's history. The Camp Bird Mine alone was one of the half dozen top mines in Colorado, and it is still producing. In the 1890's the Camp Bird made its owner, Thomas Walsh, a multi-millionaire and an internationally known figure. His phenomenal rise and sudden death were chronicled in the Colorado mining classic, *My Father Struck It Rich,* written by his daughter, Evalyn Walsh McLean, one-time owner of the jinxed Hope Diamond. Some of the other great mines burrowed into the mountain massif were the Revenue, Ruby, Virginius, Smuggler-Union, Sheridan, Mendota, and Tomboy. There is evidence that virtually every foot of the mountain has been prospected for gold and silver. Rich mining lodes were discovered high on its slopes, and prosperous mining towns such as Sneffels and Ruby once thrived way up the mountain.

Despite the pockmarks left by mining, Mount Sneffels remains one of the most beautiful and most photographed mountains in the state. The mountain massif has no less than 13 peaks which near the 14,000-foot mark; the most prominent among these are Kismet, Tea-Kettle, Potosi, Cirque, Gilpin, and Mears.

The exact origin of Mount Sneffels' name has still not been determined. F. V. Hayden originally named it "Mount Blaine" for James Blaine, a leading advocate of free silver and an unsuccessful candidate for President. However, members of Hayden's own group compared the mountain to the icelandic Mount Snaefell of Jules Verne's novel, *Mountains on the Moon.* Some claim the mountain was named for a Mr. Sneffels said to be an unlisted member of the Hayden party. Finally, it is said that early miners called the peak "Mount Sniffles" for the colds they acquired in the mines.

Mount Sneffels' many distinctive features include the fact that it offers an endless variety of climbs, from the very easy to the nearly impossible. The first recorded climb was made in 1874 by Franklin Rhoda of the Hayden Survey team, who reported that this was his first major challenge in Colorado. After spending great effort to reach the summit, Rhoda found that a grizzly bear had already been there first. The perilous north face was first officially negotiated in 1931. The first recorded climb up the northwest ridge over the prominent gendarme known as "Purgatory Point" was made by Henry McClintock and his son and daughter in 1953. As so often happened, however, prospectors, too hungry for gold and silver to worry about the danger of the climb, were probably first to reach the summit.

Despite the great wealth of the mountain, convenient, safe travel has always been a problem. Four attempts to build a toll road from Ouray to Mount Sneffels failed during the years 1876 to 1878. Even after a road was successfully built, the journey continued to be treacherous because of the danger of mud and snow slides. Today, the jeep trail leading to old mining sites is one of the most frightening mountain rides in the entire state.

WILSON PEAK (San Miguel Range) 14,017 feet
47th highest

The Fourteeners, as giants in the earth, often bear the names of important and well-known persons. However, not all of Colorado's most impressive citizens could be honored when it came to naming the state's great landmarks. Even more curious is the fact that one non-Coloradan is the namesake of two mighty Fourteeners.

Wilson Peak and Mount Wilson were both named not in honor of a President of the United States, but for A. D. Wilson, a chief topographer with the Hayden Survey. Wilson proved himself an excellent surveyor, for the Hayden teams' recorded mountain heights are remarkably close in many cases to altitudes determined later by more thorough modern measurements. Wilson was also quite a mountaineer. In 1870 he was the second man to climb Mount Rainier; his ascent of the lofty Washington peak was accomplished only two months after the first successful attempt.

The Silver Pick, one of Colorado's most celebrated mines, is at the 11,000-foot level of Wilson Peak and makes an excellent campsite for climbers. Negotiating this peak requires the better part of a day and is an excellent conditioner for the ascent up Mount Wilson, which can also be made from the Silver Pick camp.

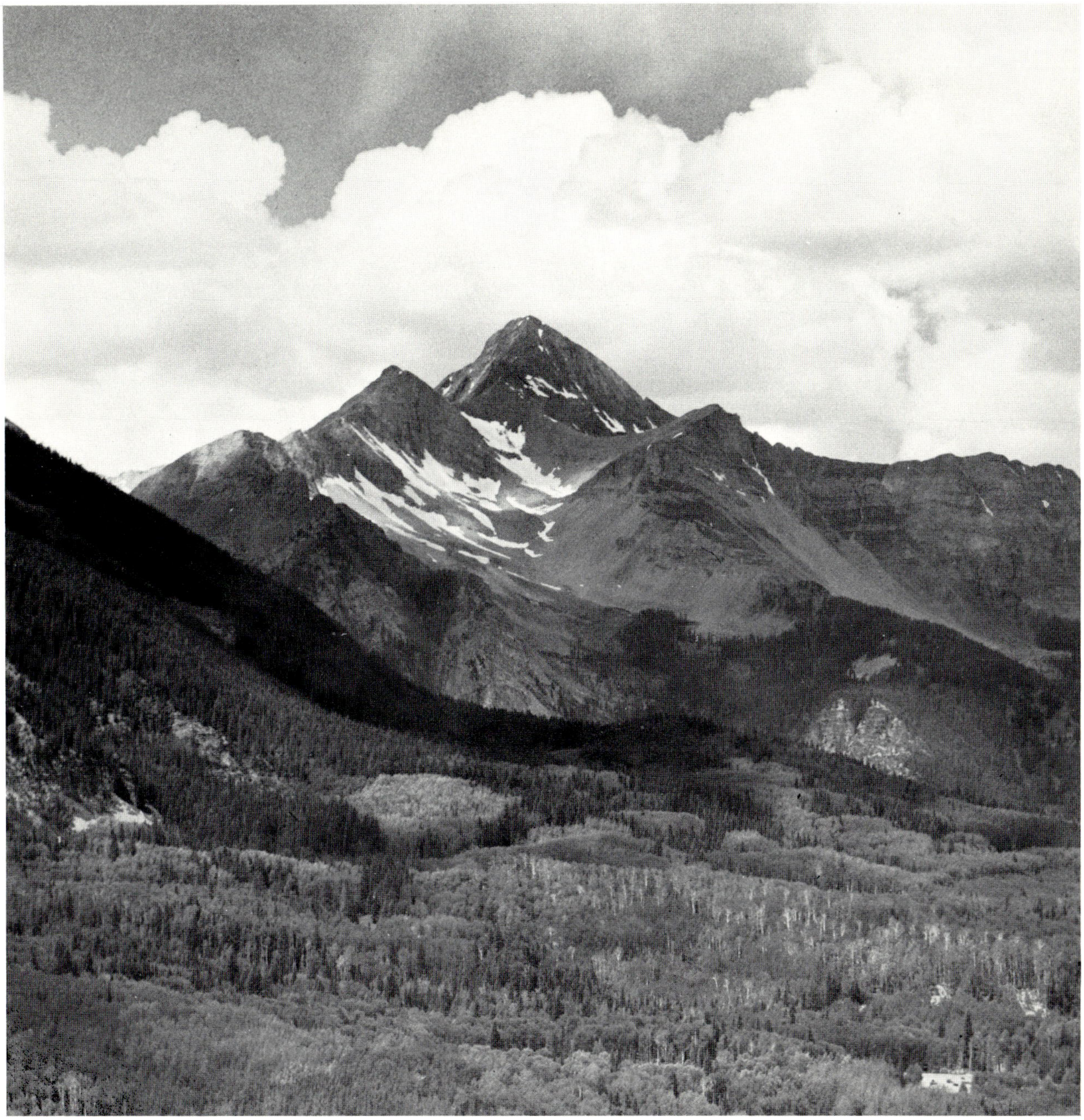

MOUNT WILSON (San Miguel Range) 14,246 feet
16th highest

A well-known alpinist once ranked Mount Wilson among the five Fourteeners most difficult to scale. Modern alpinists might argue about this, but at least in addition to its being a very interesting climb, this mountain is equal in its rugged beauty to the imposing Pikes Peak, Mount Evans, and Longs Peak.

The mountain was originally called "Glacier Peak," but U.S. Geologist F. V. Hayden later renamed it for his topographer, A. D. Wilson. It is believed that before Wilson scaled this mountain in 1874, no previous attempt had succeeded. In 1945 Colorado Governor Vivian asked the legislature to rename the mountain for President Franklin Roosevelt. The movement for change received wide support and came so close to success that some mapmakers actually began listing the mountain as "Mount Franklin Roosevelt."

Though it is a rigorous climb, Mount Wilson is one of today's most popular climbing mountains. Many experienced alpinists have scaled both Mount Wilson and its neighbor, Wilson Peak, in the same day. There are many trails up the peak, though the most favored route starts from the famed Silver Pick Mine, located at the 11,000 foot mark on Wilson Peak.

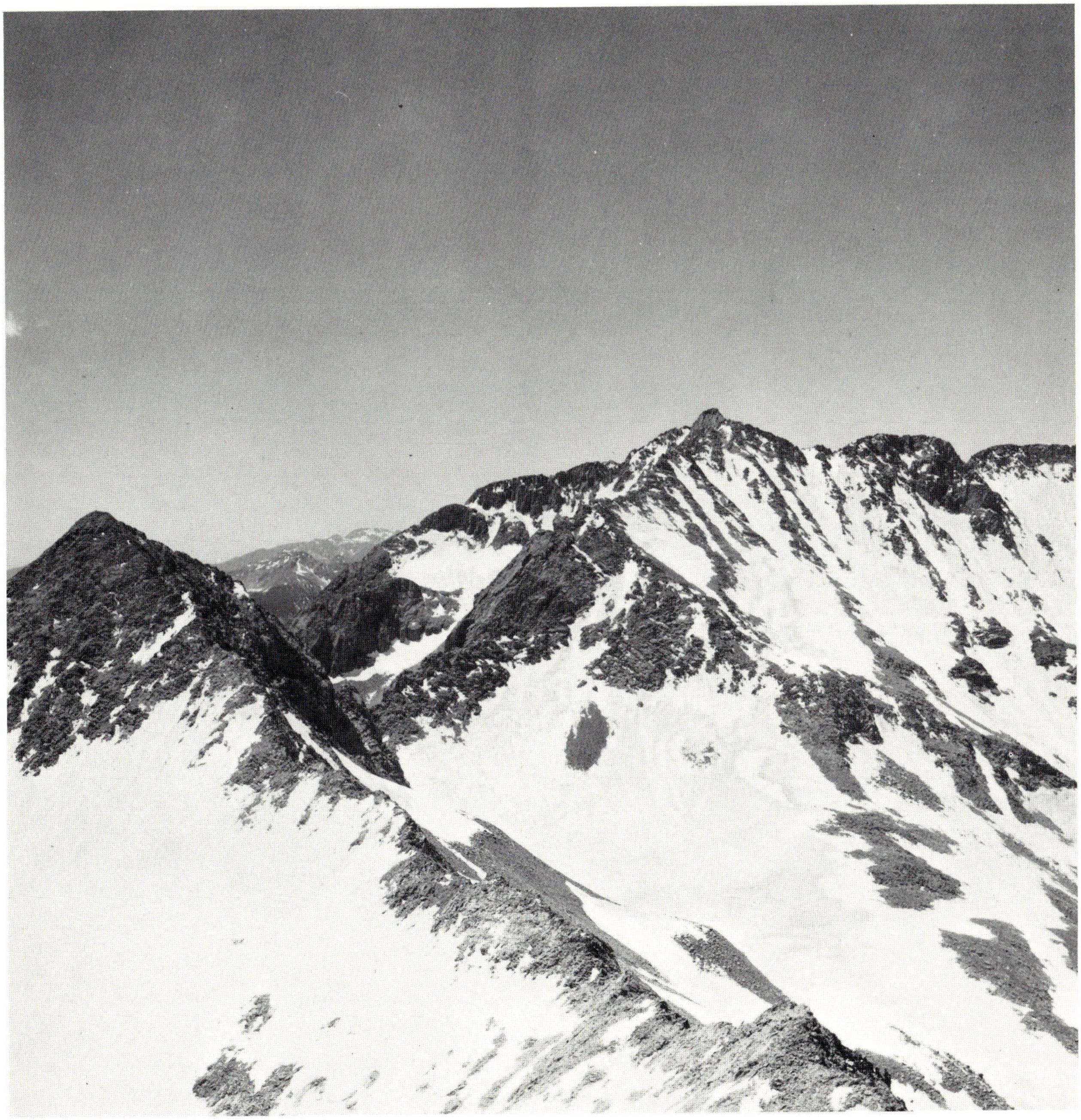

EL DIENTE PEAK
(San Miguel Range) 14,159 feet
24th highest

Jagged and steep, El Diente is considered one of Colorado's most demanding climbs. Its name means "the tooth" in Spanish, and evidently the peak was named for the ragged outline it presents. It was once believed that the first ascent up this peak did not take place until 1928, which would have made it the last Fourteener to be scaled. However, Dwight Lavender, who was among the first to conquer this peak in recent times, believes the peak was actually first scaled in 1890 by one Perry Thomas. Thomas apparently believed he had climbed neighboring Mount Wilson, but Lavender is convinced that Thomas' report of the climb leaves little doubt that the mountain he scaled was El Diente. Thomas reported leaving a cairn on top of the peak to mark his success, though the 1928 party did not discover it there. However, cairns may often be destroyed in less than a year, and the weather had nearly forty years to work on this one.

European climbers report that El Diente is one of the few Colorado mountains as challenging as the Alps. A register was placed on the mountain during the 1928 climb, and only 38 names were listed during the next 14 years. Today the peak is generally climbed from Mount Wilson or from Dunton Meadows.

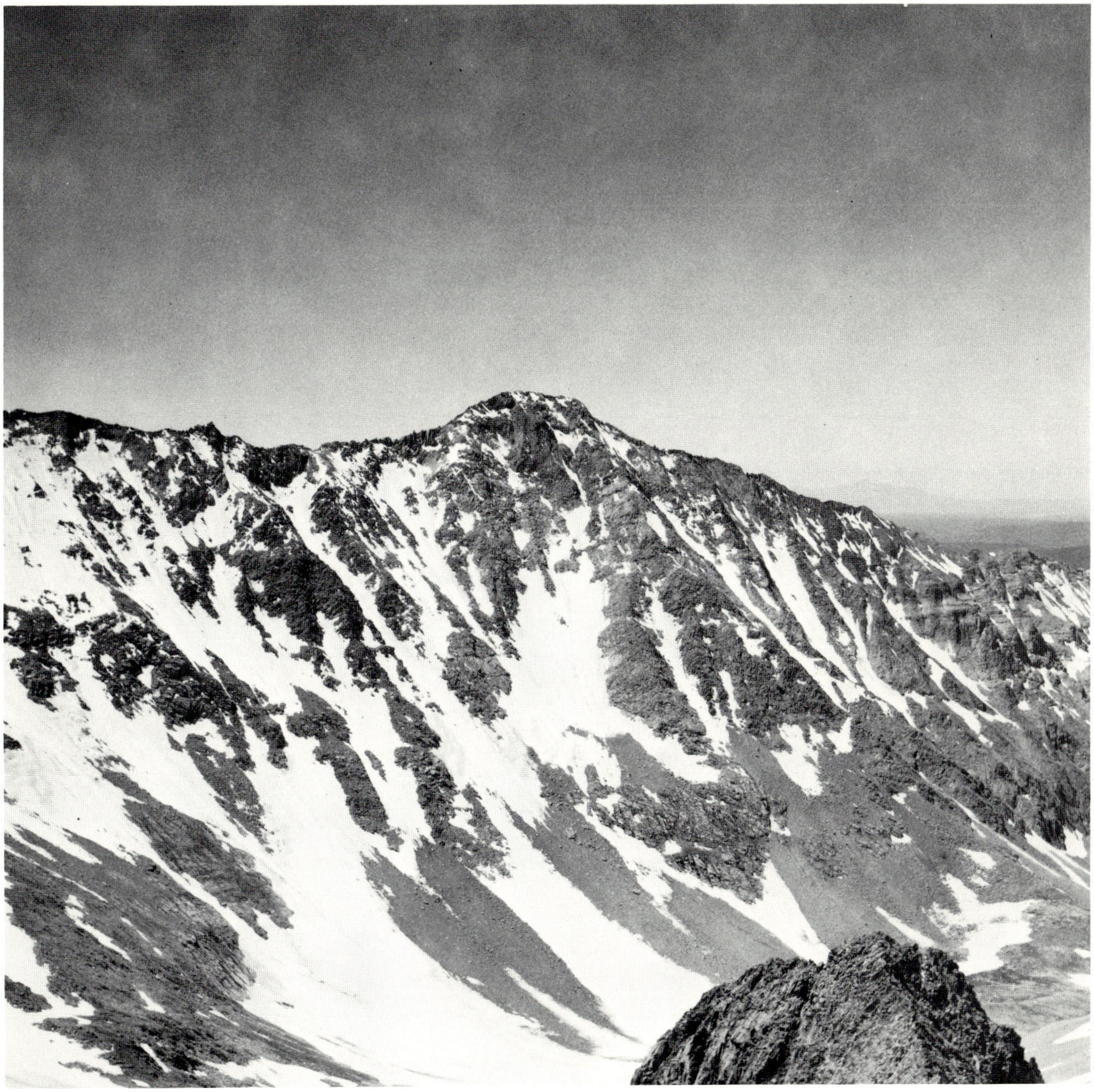

Welcome to the Club
MOUNT ELLINGWOOD
(Sangre do Cristo Range) 14,042 feet
Provisional Membership

No, they didn't make a new mountain. It's been there all along. They just got around to mapping it as a Fourteener. The newly-mapped (by the USGS) Blanca Peak quad shows this latest member of the fraternity. It is located about ½ mile northwest of Blanca and separated from it by a long, sharp saddle which drops to about 13,700 feet.

Though Ellingwood is mapped as a Fourteener and seems sufficiently separate from Mt. Blanca to meet fraternity requirements, the Colorado Mountain Club has not had the chance as yet to put it to all the proper tests and give it its sanction. "Proper tests" include verification for the height, whether a peak is far enough (¼ mile at least) from the "mother" peak to merit classification as a separate mountain, and whether the shoulder separating them drops sufficiently (300 feet at least) to make it more than just a shoulder of the mother mountain.

In one sense, the U.S. Geological Survey feels that its calculations are the final word and the Colorado Mountain Club feels that its own sanction is definitive. However, a kind of unwritten agreement has been worked out making each equally authoritative. In other words, a Fourteener is not really legitimate until adopted by both parents.

Mount Ellingwood was named for Albert R. Ellingwood, Colorado Mountain Club pioneer, who, according to his memorial in *Trail and Timberline* shortly after his death on May 12, 1934, "did more for Colorado Mountaineering than any other man". He was one of the first three men to climb all of the Fourteeners; the first to climb Crestone Needle, Crestone Peak, and Kit Carson Peak (in 1916), Lizzard Head, Pigeon Peak, and Turret Peak (in 1920); the mountaineer to introduce in Colorado proper rope climbing techniques that made theretofore "impossible" climbs possible and helped bring a measure of safety to mountain climbing. He also broke many new trails up the mountains, including two bearing his name: Ellingwood Ridge on La Plata Peak and Ellingwood Arete on Crestone Needle.

Professor Ellingwood graduated from Colorado College and won a Rhodes Scholarship to Oxford. He taught political science at Colorado College and later at Lake Forest College and Northwestern University in Illinois. Despite his many educational, administrative, and writing tasks, he spent every spare moment climbing Colorado and Wyoming peaks, doing more in his short 46 years than most men accomplish in twice the time.

The same new U.S.G.S. maps of the Sangre de Cristo Range which welcomed Mount Ellingwood to the club also revised elevation figures for at least four other Fourteeners.

See how dynamic mountains are?

Honorable Mention
STEWART PEAK (San Juan Range)

Stewart was only recently booted out of the Fourteener fraternity. Once recorded as the 42nd highest, a 1958 U.S.G.S. survey trimmed the mountain down to 13,983 feet — a loss of 77 feet. It then took nearly ten years to make the change official.

The peak was named by the Wheeler Survey in the 1870's for Nevada's U.S. senator William M. Stewart, a prominent figure in the Comstock Lode controversy and a leading advocate of free silver.

MOUNT CAMERON (Mosquito Range)

This peak, which measures 14,238 feet, was once considered one of Colorado's highest mountains. In recent years, however, the U.S.G.S. and the Colorado Mountain Club have agreed that it is too much a part of Mount Lincoln to deserve separate identity as a Fourteener. It is often called Cameron Peak.

GRIZZLY MOUNTAIN (Sawatch Range)

Grizzly Mountain, or Peak as sometimes called, is located south of Independence Pass. It was a Fourteener until the 1965 U.S.G.S. listing knocked five or six feet off its official height, leaving it just under 14,000 feet.

NORTH EOLUS PEAK (Needles Range)

This is a case where the U.S.G.S. and the Colorado Mountain Club differ. The former classifies North Eolus as a Fourteener, while the Colorado Mountain Club does not. Disagreement is over shoulder drop.